STRENGTHENING THE GRID

Effect of High-Temperature Superconducting Power Technologies on Reliability, Power Transfer Capacity, and Energy Use

Richard Silberglitt ◆ Emile Ettedgui ◆ Anders Hove

Science and Technology Policy Institute

RAND

Prepared for the Department of Energy

The research described in this report was conducted by RAND's Science and Technology Policy Institute for the Department of Energy under contract ENG-9812731.

Library of Congress Cataloging-in-Publication Data

Silberglitt, R. S. (Richard S.)
    Strengthening the grid : effect of high temperature superconducting (HTS) power
technologies on reliability, power transfer capacity, and energy use / Richard Silberglitt,
Emile Ettedgui, and Anders Hove.
        p. cm.
    Includes bibliographical references.
    "MR-1531."
    ISBN 0-8330-3173-2
    1. Electric power systems—Materials. 2. Electric power systems—Reliability. 3. High
temperature superconductors. I. Ettedgui, Emile. II. Hove, Anders. III.Title.

TK1005 .S496 2002
621.31—dc21

                                                                                2002021398

Published 2002 by RAND
1700 Main Street, P.O. Box 2138, Santa Monica, CA 90407-2138
1200 South Hayes Street, Arlington, VA 22202-5050
201 North Craig Street, Suite 102, Pittsburgh, PA 15213
RAND URL: http://www.rand.org/
To order RAND documents or to obtain additional information, contact Distribution Services: Telephone: (310) 451-7002; Fax: (310) 451-6915; Email: order@rand.org

Preface

Power transmission constraints resulting from the slow growth of transmission systems relative to the large growth in the demand for power have played a major role in higher electricity prices and reduced reliability in many areas across the United States in recent years. This report summarizes the results of two RAND studies that investigated the potential effect of high-temperature superconducting (HTS) technologies on the U.S. power grid.

In the first study, which was supported by the U.S. Department of Energy (DOE) and performed under the auspices of RAND's Science and Technology Policy Institute, RAND reviewed the evolution of the U.S. electricity market under deregulation, identified electricity transmission constraints, and made engineering comparisons between conventional power components and those based upon HTS technologies. That study was begun in July 2000 and concluded with this report. In the second study, which was supported by the National Renewable Energy Laboratory, RAND analyzed the effects of HTS transmission cables in realistic grid situations using power-flow simulations performed for RAND by PowerWorld Corporation. HTS cable manufacturer Pirelli Cables and Systems provided engineering data required for these simulations.

The second study was conducted from June 2001 to January 2002. This report integrates the results of those two studies to draw conclusions concerning ways in which HTS cables could be deployed in the future to increase grid reliability and power transfer capacity. This report also compares the energy use and acquisition and operating costs of HTS and conventional power technologies.

This report should be of interest to individuals and organizations that generate, transmit, distribute, use, or regulate electric power, all branches of government that deal with electric power systems, and individuals and organizations that sponsor or perform research and development on superconducting and related technologies.

Originally created by Congress in 1991 as the Critical Technologies Institute and renamed in 1998, the Science and Technology Policy Institute is a federally funded research and development center sponsored by the National Science Foundation and managed by RAND. The Institute's mission is to help improve public policy by conducting objective, independent research and analysis on policy issues that involve science and technology. To this end, the Science and Technology Policy Institute

- supports the Office of Science and Technology Policy and other Executive Branch agencies, offices, and councils
- helps science and technology decisionmakers understand the likely consequences of their decisions and choose among alternative policies
- helps improve understanding in both the public and private sectors of the ways in which science and technology can better serve national objectives.

Science and Technology Policy Institute research focuses on problems of science and technology policy that involve multiple agencies. In carrying out its mission, the Institute consults broadly with representatives from private industry, institutions of higher education, and other nonprofit institutions.

Inquiries regarding the Science and Technology Policy Institute may be directed to:

Helga Rippen
Director
Science and Technology Policy Institute
RAND
1200 South Hayes Street
Arlington, VA 22202-5050
Phone: (703) 413-1100 x5574
Email: http://www.rand.org/scitech/stpi/
Web: http://www.rand.org/centers/stpi

Contents

vi

Figures

Tables

Summary

This report evaluates the potential of high-temperature superconducting (HTS) power technologies to address existing problems with the U.S. electric power transmission grid, especially problems with transmission constraints. These constraints that have resulted from the slow growth of transmission systems relative to the growth in demand for power have played a major role in higher electricity prices and reduced reliability in a number of areas across the United States in recent years. Electric power components using superconducting materials have the potential to address these transmission constraints because they have much higher energy density than conventional power equipment, which for transmission means added power-carrying capacity.

Superconducting power equipment requires cooling to sustain operating temperatures hundreds of degrees below ambient temperature. Magnets based on low-temperature superconducting (LTS) materials that require cooling with liquid or gaseous helium have become commercial products for accelerator and magnetic resonance imaging applications. However, the cost of cooling these LTS materials is a substantial barrier to their use in power system components. HTS power equipment, on the other hand, can be cooled with liquid nitrogen which is considerably cheaper than liquid or gaseous helium, thereby reducing or eliminating this cost barrier.

Projects are underway throughout the world to demonstrate the following superconducting electric power components:

- Low-loss and high-capacity power transmission cables

- Compact, high-efficiency, and low-environmental-impact transformers

- Storage of electricity via persistent currents in a coil (superconducting magnetic energy storage or SMES) or persistent rotation with a magnetic bearing (flywheel energy storage systems or FESS).

The various HTS technologies are at different stages of development: commercial SMES, composed of magnets with LTS wire and HTS current leads, has begun to appear in niche markets. HTS cables have been demonstrated at full scale at distribution voltages and in lengths up to a hundred meters. HTS FESS and transformers are also being demonstrated, but at less than full scale.

xii

Computer simulations described in the body of this report show that HTS transmission cables produce a beneficial redistribution of power flows when substituted for conventional cables or conductors or added to the transmission grid. Increased power flows through the HTS cables can relieve stress on a heavily loaded network, resulting in increased reliability and increased power transfer capacity at the same level of reliability. This capability for downtown Chicago is demonstrated in this report. HTS cables can also provide an alternate transmission path at lower voltage to increase transfer capacity. This is demonstrated for California's Path 15 (the northern portion of the grid link between northern and southern California). The length of the simulated HTS cables is consistent with near-term demonstrations for Chicago, but is much longer than what is possible in the near term for Path 15.

At what cost do HTS cables provide these benefits? Engineering comparisons of HTS and conventional power components described in this report identify the range of parameters, principally component utilization and cooling efficiency, for which HTS power components may use less energy than conventional power components. Lower energy use translates into lower operating cost. These engineering comparisons also identify the range of HTS component (plus cooling system) costs and electricity costs in order for the operating cost reduction to be greater than the higher HTS component acquisition cost.

We draw the following conclusions from the data presented and analysis described in this report:

1. Significant transmission constraints exist in many areas of the United States. These constraints have resulted from increased demand, increased power transfers, and very small increases in transmission capacity over the past several years.

2. These transmission constraints have contributed, in some cases, to decreased reliability and to price differentials between load areas.

3. HTS underground cables provide an attractive retrofit option for urban areas that have existing underground transmission circuits and wish to avoid the expense of new excavation to increase capacity. This situation exists because HTS cables have almost zero resistance, very small capacitance and inductance, and high power capacity compared with conventional cables of the same voltage. Thus, the HTS cables provide changes in power flows that reduce stress in heavily loaded circuits, thereby increasing reliability or power-transfer capacity and relieving transmission constraints.

4. When operated at high utilization, HTS cables provide energy savings benefits compared with conventional cables per unit of power delivered for a

range of HTS cable parameters consistent with existing data and engineering estimates. However, whether or not the concomitant HTS cable operating cost savings are greater than the increase in acquisition cost compared with conventional cables depends on the cost of electricity.

5. HTS cables can provide a parallel transmission path at lower voltage to relieve high-voltage transmission constraints. The implementation of this approach for long-distance transmission circuits will depend on the development of periodic cooling stations and sufficient manufacturing capacity for HTS wire.

6. When operated at high utilization, HTS cables provide energy savings benefits compared with conventional overhead lines per unit of power delivered for a range of HTS cable parameters consistent with existing data and engineering estimates. HTS cables may also provide concomitant life-cycle cost benefits for situations in which the usual cost advantage of overhead transmission lines is mitigated by site-specific concerns such as high land-use demands or right-of-way costs or the expense of obtaining siting approvals or increasing power transfer capacity at higher voltage.

7. Flywheel energy storage systems using HTS magnetic bearings are in the demonstration stage and have the potential to achieve performance characteristics that will make flywheels competitive with batteries in a wide range of electricity storage applications.

8. HTS transformers can provide increased power capacity with the same footprint as conventional transformers and could be sited inside buildings because they eliminate fire hazards associated with oil. If estimated HTS-wire cost reductions from a new manufacturing facility are achieved and the wire meets performance requirements, HTS transformers are projected to be cost competitive with conventional transformers for a range of parameters consistent with existing data and engineering estimates.

9. Superconducting magnetic energy storage systems that use low-temperature superconducting coils and HTS current leads have already found a niche market for distributed reactive power support to prevent grid voltage collapse and for maintaining power quality in manufacturing facilities.

The White House Office of Science and Technology Policy (OSTP) authored a National Action Plan on Superconductivity Research and Development more than a decade ago. Based upon the analysis described in this report, a 2002 action plan for HTS power technologies might include the following recommendations:

• The DOE-led HTS power technologies research and development (R&D) program should continue to emphasize second-generation wire development, with the goal of providing HTS wire meeting commercial cost and performance targets.

- The DOE Superconductivity Partnership Initiative should be expanded, building on the current demonstrations that are providing operating experience, to develop new demonstrations with operational energy or power transfer benefits. These new demonstrations should include HTS cable demonstrations at longer lengths and transmission voltages and demonstrations of HTS transformers and FESS at a scale consistent with utility substation and end-user facility needs.

- The HTS R&D program should increase emphasis on and support for the development of cryocoolers with multi-kilowatt capacity that can be mass produced. These cryocoolers should have an efficiency rating greater than today's range of 14-20 percent of the highest possible (Carnot) efficiency at HTS-power-component operating temperatures.

- The HTS R&D program should increase the emphasis on and support for the development of standards for HTS-power-component installation and operation and increase the emphasis on and support for training of industrial staff who will operate and maintain these installations.

Acknowledgments

The authors express their appreciation to Alan Wolsky of Argonne National Laboratory and Martin Libicki of RAND for their detailed and insightful review of the draft of this report, which contributed significantly to both its structure and content. The authors are also grateful to Bruce Don and Stephen Rattien of RAND, Anthony Schaffhauser of the National Renewable Energy Laboratory, Robert Hawsey of Oak Ridge National Laboratory, James Daley of the Department of Energy, and Terrence Kelly, formerly of the OSTP and now with RAND, for their support in the initiation of this study.

The authors acknowledge the assistance of many individuals in the electric utility, power engineering, and superconductivity R&D communities without whose willingness to share data, observations, knowledge, and insights this study could not have been successfully completed. Those individuals include:

Paola Caracino (Pirelli), Michel Coevoet (Electricité de France), Steinar Dale (ABB), Arthur Day (Boeing), Larry Goldstein (National Renewable Energy Laboratory), Paul Grant (EPRI), Santiago Grijalva (PowerWorld), Richard Hockney (Beacon Power), Nathan Kelley (Pirelli), Albert Keri (AEP), Brendan Kirby (Oak Ridge National Laboratory), Mark Laufenberg (PowerWorld), Chet Lyons (American Superconductor), Andrea Mansoldo (Pirelli), Ben McConnell (Oak Ridge National Laboratory), Sam Mehta (Waukesha), Joseph Mulholland (Arizona Power Authority), Martin Nisenoff (Naval Research Laboratory, retired), Thomas Overbye (University of Illinois), Marshall Reed (Department of Energy), Thomas Sheahen (Western Technology), Michael Strasik (Boeing), Christopher Wakefield (Southern Company), Alan Wolsky (Argonne National Laboratory), John Ziegler (Houston Advanced Research Center), and several anonymous Oak Ridge National Laboratory reviewers of interim draft reports.

Finally, the authors acknowledge the assistance in the final preparation of this report of RAND colleagues Debra Knopman, whose suggestions directed the report toward a broader audience, and Nancy DelFavero, whose insightful editing contributed significantly to the clarity of the text.

Acronyms

ABB	Asea Brown Boveri (former name)
AC	Alternating current
ACSR	Aluminum conductor steel reinforced
AEP	American Electric Power
Al	Aluminum
A	Amperes
BESS	Battery energy storage system
BSCCO	Bismuth strontium calcium copper oxide
CBEMA	Computer and Business Equipment Manufacturers Association
CSA	Corrugated seamless aluminum
Cu	Copper
DC	Direct current
Delmarva	Delaware-Maryland-Virginia
DOE	Department of Energy
ECAR	East Central Area Reliability Coordination Agreement
EHV	Extra-high voltage
EIA	Energy Information Administration
EPACT	Energy Policy Act of 1992
EPRI	Electric Power Research Institute (former name)
ERCOT	Electric Reliability Council of Texas
FACTS	Flexible AC transmission system
FERC	Federal Energy Regulatory Commission
FESS	Flywheel energy storage system
FOIA	Freedom of Information Act
FRCC	Florida Reliability Coordinating Council
GW	Gigawatt(s)

HTS	High-temperature superconducting
HV	High voltage
HVDC	High-voltage direct current
IEEE	Institute of Electrical and Electronics Engineers
IGC	Intermagnetics General Corporation
ISO	Independent system operator
ITI	Information Technology Industries Council
kA	Kiloamperes
kAm	Kiloamperes times meters
km	Kilometer(s)
kV	Kilovolt(s)
kVA	Kilovolt(s) amperes
kWh	Kilowatt hour(s)
LTS	Low-temperature superconducting
m	Meter(s)
MAAC	Mid-Atlantic Area Council
MAIN	Mid-America Interconnected Network
MAPP	Mid-Continent Area Power Pool
MJ	Megajoules
MVA	Megavolt amperes
MVAR	Megavolt amperes reactive
MW	Megawatt(s)
mW	Milliwatt(s)
NASA	National Aeronautics and Space Administration
NERC	North American Electric Reliability Council
NIST	National Institute of Standards and Technology
NPCC	Northeast Power Coordinating Council
OSTP	Office of Science and Technology Policy
PG&E	Pacific Gas and Electric Corporation
PJM	Pennsylvania–New Jersey–Maryland

POST	Power Outage Study Team
PURPA	Public Utilities Regulatory Policy Act
R&D	Research and development
RTO	Regional transmission organization
SERC	Southeastern Electric Reliability Council
SMES	Superconducting magnetic energy storage
SPI	Superconductivity Partnership Initiative
SPP	Southwest Power Pool
TLR	Transmission-loading relief
TVA	Tennessee Valley Authority
UPS	Uninterruptible power supply
VAR	Volts amperes reactive
W	Watt(s)
W/kAm	Watts per kiloampere meter
W/kg	Watts per kilogram
Wh/kg	Watt hours per kilogram
W/m	Watts per meter
WSCC	Western Systems Coordination Council
XLPE	Cross-linked polyethylene
YBCO	Yttrium barium copper oxide

Symbols

A_c	Acquisition cost of cable or conductor
A_r	Acquisition cost of refrigeration
A_t	Acquisition cost of transformer
c	Transformer core loss
$C_{r/i}$	Conversion efficiency of rectifier/inverter
C_s	Conversion efficiency of energy storage module
d	Discount factor
I	Cable or conductor current
I_L	Idling loss of FESS or BESS
$\bar{i}_F$	Idling loss of FESS per unit output power
l	Length of cable or conductor
O_c	Contribution of conduction to the operating cost of cable or conductor
O_r	Contribution of refrigeration to the operating cost of cable or conductor
P	Power dissipated in cable, conductor, or transformer
P_c	Power dissipated through conduction
P_L	Power dissipated in energy storage device
P_r	Power used for refrigeration
R	Power ratio
T_c	Cold (operating) temperature of superconducting power equipment
T_h	Hot (ambient) temperature of superconducting power equipment
u	Utilization
ε	Cost of electricity
η_C	Carnot efficiency
η	Cryocooler efficiency as a fraction of Carnot efficiency

θ	Thermal loss per unit length of HTS cable or transformer
ρ	Cryocooler coefficient of performance
τ	Termination loss of HTS cable
χ	Cable or conductor cost
ω	Conduction loss per unit length of cable, conductor, or transformer

1. Introduction

The continuing growth in demand for electric power, coupled with deregulation and restructuring of electricity markets in recent years, has placed increased demands on the electricity transmission and distribution grid in many areas of the United States. Whereas formerly regulated utilities kept substantial reserve margins for both generation and transmission within well-defined service areas, the current environment, with its increasing amount of power produced and sold in competitive markets, is characterized by increased power transfers and lower reserve margins, leading to increased difficulty in maintaining reliability.[1]

The current market environment also provides much greater incentives for new generation than for new transmission. Data available from the U.S. Energy Information Administration (EIA) and the North American Electric Reliability Council (NERC) show that the ratio of transmission capacity to peak electricity demand in the United States has decreased by 16 percent between 1989 and 1998, and is projected to decrease by another 12 percent by 2008.[2]

The need for increased transmission investment has been recognized by the Federal Energy Regulatory Commission (FERC), which has also sought to increase regional coordination of transmission planning.[3] The 2001 National Energy Policy Report notes that transmission constraints have played a role in higher electricity prices and reduced reliability, identifies several locations in the U.S. at which transmission congestion is a serious problem, and includes among its recommendations expanded research and development (R&D) on transmission reliability and superconductivity.[4] Superconductivity is of interest here because superconductors transmit electricity with a small fraction of the losses from conventional conductors, which enables power transmission at higher power densities, which in turn could relieve transmission congestion.

---

[1]Mountford, John D., and Ricardo R. Austria, "Keeping the Lights On," *IEEE Spectrum,* June 1999, pp. 34–39.

[2]Hirst, Eric, "Expanding U.S. Transmission Capacity," Oak Ridge, Tenn., August 2000, p. 58, http://www.eei.org/future/reliability/hirst2.pdf (last accessed April 1, 2002).

[3]Federal Energy Regulatory Commission, *Regional Transmission Organizations, Order No. 2000,* Washington, D.C.: Federal Energy Regulatory Commission, Docket No. RM99-2-000, December 20, 1999.

[4]National Energy Policy Development Group, *National Energy Policy,* U.S. Government Printing Office, May 2001, pp. 7-5 to 7-6.

2

From the discovery of superconductivity by Kammerlingh Onnes in 1911 until
the discovery of high-temperature superconducting (HTS) materials by Bednorz
and Muller in 1986, superconductivity was a phenomenon that occurred only
below 25 K (–451°F),[5] which required refrigerators operating at cryogenic
temperatures, or *cryocoolers*, using liquid and gaseous helium. A number of
electric power components—including motors, generators, and both alternating
current (AC) and direct current (DC) power cables—were conceived and
developed beginning in the 1970s that used the low-temperature
superconducting (LTS) materials Nb_3Sn, with a superconducting transition
temperature of 18 K, and NbTi, with a superconducting transition temperature of
10 K. While these projects provided useful engineering and design data, they did
not lead to commercial products, in part because of cost considerations, in part
because of technical problems, and especially in the case of motors and
generators, in part because of the trend away from very large central power-
generating facilities.[6]

Power equipment that uses HTS materials (T_c>77 K) can be cooled with
cryocoolers based upon liquid nitrogen, which is considerably cheaper than
liquid helium. However, the cost of the HTS wire is currently much higher than
the cost of conventional conductors used for electricity transmission and
distribution, such as copper (Cu) or aluminum (Al). Wire cost is measured in
dollars per thousand amperes (kA) of current-carrying capacity times length
measured in meters (m). Today's commercial HTS wire, made of $Bi_2Sr_2Ca_2Cu_3O_{10}$,
or bismuth strontium calcium copper oxide (BSCCO), costs $200 per kAm
(kiloamperes times meters),[7] compared with $9.56 per kAm for Cu and $1.60 per
kAm for Al.[8] The DOE Superconductivity Program for Electric Systems has
established a cost goal of $10 per kAm for second-generation HTS wire
(envisioned as a $YBa_2Cu_3O_7$ or yttrium barium copper oxide [YBCO], coating on a
conducting substrate).

Aside from potential benefits of reducing transmission congestion or increasing
reliability, the value of HTS power equipment will be in its ability to improve the
efficiency, and thus reduce the cost, of systems for electricity production,

[5]Golubov, A. A., "High Temperature Superconductivity," *Handbook of Applied Superconductivity*, Vol. 1, Bristol, UK, and Philadelphia: Institute of Physics Publishing,, 1998, pp. 53–62.

[6]Grant, Paul M., "Superconductivity and Electric Power: Promises, Promises...Past, Present and Future," *IEEE Transactions on Applied Superconductivity*, Vol. 7, No. 2, June 1997, pp. 112–133.

[7]Lyons, Chet, American Superconductor, private communication. John Howe, Vice President, Electric Industry Affairs, American Superconductor, projected that following a scale-up of its manufacturing facility (planned for 2004), American Superconductor will be able to offer BSCCO at $50 per kAm (briefing on superconductivity at Rayburn House Office Building, August 3, 2001).

[8]These estimates are based upon current densities in conventional transmission cables and conductors analyzed in Chapter 4.

conversion, transmission, and use. A recent study sponsored by the DOE developed a methodology for estimating the market penetration of HTS cables, transformers, motors and generators, based upon the value of energy efficiency improvements, in terms of engineering parameters including the cost of HTS wire. Initial estimates based on current values and estimates of the parameters suggest significant market penetration at a wire cost of $20 per kAm,[9] although the authors of that study clearly indicate that their report is intended to provide a framework for improved estimation as engineering parameters become better defined.

This RAND report does not directly address HTS market issues. Rather, it focuses on two different one-to-one comparisons between HTS and conventional power technologies. First, we evaluate the effects of HTS cables within an interconnected network of conventional power equipment on reliability, as measured by the number and severity of overloaded transmission lines resulting from single line outages, and on power transfer capacity, the quantity most closely related to transmission congestion. Second, we compare the energy use, acquisition cost, and operating cost of HTS and conventional power equipment per unit of delivered power, and use the cost results to estimate, in terms of the engineering parameters of the HTS technologies, the time for the higher HTS acquisition cost to be recovered through operating cost savings.

Demonstration projects are underway throughout the world to use the unique properties of superconductors (e.g., low loss, high energy density) in electric power components--for example, storage of electricity via persistent currents in a coil (superconducting magnetic energy storage or SMES) or persistent rotation with a magnetic bearing (flywheel energy storage system or FESS); low-loss and high-capacity power transmission cables; compact, high-efficiency, and low-environmental-impact transformers. The capabilities of these technologies, together with the capabilities of new and evolving conventional technologies, such as flexible AC transmission systems and high-voltage DC transmission that allow control over the direction of power transfers, may provide the building blocks for future grid improvements to address transmission congestion. For example, the low-loss and high-power capacity of HTS cables could be used to modify power-flow paths to reduce the loading on existing transmission lines, which would increase either reliability (as defined in the previous paragraph) or power transfer capacity.

[9]Mulholland, Joseph, Thomas P. Sheahen, and Ben McConnell, *Analysis of Future Prices and Markets for High Temperature Superconductors* (draft), U.S. Department of Energy, September, 2001, http://www.ornl.gov/HTSC/pdf/Mulholland%20Report.pdf (last accessed April 1, 2002).

Superconducting power technologies also have characteristics that may enable them to contribute to reducing certain vulnerabilities of the power grid. HTS cables will be sited underground, making the cables themselves (like conventional underground cables) less vulnerable to natural events and unintentional or intentional disruptions than overhead transmission lines. HTS cables differ from conventional cables in two ways that are relevant to a vulnerability analysis:

- First, HTS cables can carry more power at the same voltage than can conventional cables. This ability to carry more power could increase vulnerability in the event of an outage of the HTS cable. However, it could also allow power transmission at a lower voltage, which could eliminate transformers and other power equipment that could fail or be targeted by saboteurs, thereby decreasing vulnerability. Thus, while the effect on vulnerability depends upon the engineering design and implementation, HTS cables introduce new capabilities that may be used to design for specified levels of vulnerability.

- Second, HTS cables are dependent on above-ground cryocoolers to maintain their operating temperature, and these cryocoolers could be a source of vulnerability, either through accidents or sabotage or through neglected maintenance. However, current demonstration systems are designed to have reserve coolant reservoirs that provide substantial response time in case of cryocooler failure, thus reducing this aspect of vulnerability. And the fact that HTS cables operate at cryogenic temperatures also makes them impervious to the temperature-related faults that are the most common failure mode for conventional transmission lines. Finally, increases in reliability or transfer capacity obtained by insertion of HTS cables that relieve stress on the system may also decrease system vulnerabilities.

Other HTS power technologies also have vulnerability benefits. HTS transformers are compact and contain no flammable liquids, which increase flexibility for siting in less-vulnerable locations (for example, within buildings). The superconducting energy-storage technologies (like the conventional storage methods such as batteries) can be used to reduce stress on heavily loaded portions of the transmission and distribution network, increasing reliability and improving power quality, which also reduces vulnerability to as well as the potential impact of events such as accidents, natural disasters, and sabotage.

The various HTS technologies are at different stages of development: commercial SMES, composed of magnets with LTS wire and HTS current leads, has begun to appear in niche markets.[10] HTS cables have been demonstrated at full scale at distribution voltages and in lengths up to 100 meters.[11] HTS FESS and transformer demonstrations at less than full scale are being completed and new full-scale demonstration projects have recently been announced.[12] The technical characteristics and demonstration status of selected electric power component developments based on HTS materials are described in Appendix A.

It is important to note that the commercial applications of superconductivity that are envisioned do not require the development of fundamentally new types of products. Rather, they substitute superconductors for other materials in existing products in order to improve their performance. Surely, this approach calls for some redesign and modification of existing products, but at least the underlying design strategy is already known.

Chapter 2 of this report reviews transmission and distribution needs resulting from recent load growth and developments in electricity deregulation and market restructuring. The discussion addresses needs based upon increased utilization of the grid, demands from increased marketing of power, and identified but unfulfilled transmission requirements, all of which lead to major U.S. transmission constraints.

Chapter 3 reports the results of computer-generated simulations of power flows with and without HTS cables in two of the transmission-constraint areas identified in Chapter 2: downtown Chicago and California's Path 15 transmission circuit. These computer-generated simulation results quantify the potential increases in reliability and power-transfer capacity that would be possible if HTS transmission cables were available to address these types of transmission congestion problems.

Chapter 4 reports the results of parametric analyses of HTS cables as compared with conventional cables and overhead transmission line conductors, given the properties of demonstrated HTS systems and a range of engineering estimates that are based on data reported in the HTS literature. This analysis identifies the required HTS wire cost and cryocooler cost and efficiency for HTS operating cost

[10]See product descriptions and press releases available at the American Superconductor website, http://www.amsuper.com.

[11]Chang, Kenneth, "High Temperature Superconductors Find a Variety of Uses," *New York Times*, May 29, 2001.

[12]"High Temperature Superconductivity, Bringing New Power to Electricity," news update, Bob Lawrence and Associates, Inc., September 26, 2001, http://www.eren.doe.gov/superconductivity/library_bulletins.html (last accessed April 1, 2002).

savings to be greater than the acquisition cost difference between the HTS and conventional transmission options. Chapter 5 provides similar analyses for FESS using HTS bearings, HTS transformers, and (LTS) SMES using HTS current leads.

Finally, Chapter 6 summarizes the conclusions drawn from the computer-generated power-flow simulations of HTS cables and the parametric analyses of HTS and conventional technologies. It concludes with a review of the recommendations from the 1990 Office of Science and Technology Policy (OSTP) action plan for superconductivity R&D and suggestions for an updated set of recommendations for a 2002 action plan based upon the results of the analyses described in Chapters 3 through 5.

2. The Evolution of the U.S. Electricity Transmission Grid: Transmission Constraints

This chapter summarizes transmission constraints resulting from the increased utilization of the U.S. electricity transmission grid and from the recent increased marketing of power. The increase in the marketing of power, in particular, requires power transfers of a size and number well beyond those envisioned by the designers of the current grid. HTS power technologies have properties that could be used to address these transmission constraints. In particular, HTS cables offer a number of benefits:

- Several times more power capacity, through existing cable conduits, than with conventional underground cables

- Greater power transfer at lower voltages than with either conventional underground cables or overhead transmission lines, allowing reinforcement of 115 kilovolt (kV) and 138 kV transmission circuits and elimination of the need for 500 kV and higher voltage transmission that requires expensive power equipment

- Lower susceptibility to temperature-related faults than with overhead lines because HTS cables are operated at cryogenic temperatures.

Other HTS power technologies also have potential grid benefits. As described in Appendix A, compared with conventional transformers, HTS transformers can provide increased power capacity with the same footprint to either optimize use of space within substations or to avoid new construction. The lack of flammable oil in HTS transformers cooled by liquid nitrogen also increases siting flexibility, which could be used to reduce vulnerability (e.g., via indoor or enclosed siting). Flywheel energy storage systems using HTS bearings and (LTS) superconducting magnetic energy storage systems that use HTS current leads can provide options for improving power quality and in some cases reduce the urgency of transmission investments.

8

Electricity Market Restructuring

A dramatic restructuring of the U.S. electricity industry is underway. As illustrated in Figure 2.1, 24 states and the District of Columbia have enacted restructuring legislation or issued a comprehensive regulatory order (although, as of this writing, restructuring was delayed in seven of those states and retail access was suspended in California).

Whereas the 1978 Public Utilities Regulatory Policy Act (PURPA) allowed power generation for the grid by non-utilities, electricity was sold at regulated rates until this restriction was removed by the Energy Policy Act (EPACT) of 1992. However, the strong impetus for deregulation and electricity market restructuring was provided by FERC, beginning with the 1996 implementation of FERC Orders 888 and 889, which required utilities with transmission lines to provide access to transmission for power transfers between other utilities and

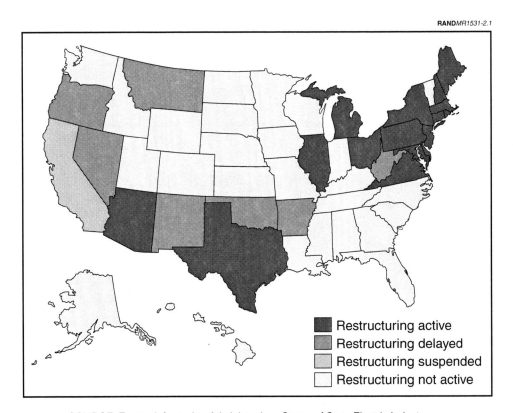

RANDMR1531-2.1

Restructuring active
Restructuring delayed
Restructuring suspended
Restructuring not active

SOURCE: Energy Information Administration, *Status of State Electric Industry Restructuring Activity*, as of March 2002, current status available at http://www.eia.doe.gov/cneaf/electricity/chg_str/regmap.html.

NOTE: The states with active restructuring legislation include the District of Columbia.

Figure 2.1—Status of Electricity Market Restructuring

non-utilities.[1] As a result of these orders, the non-utility share of generation has risen dramatically, and power transfers across regions have grown in size and number.[2]

In the past few years, FERC has pushed utilities to separate their transmission and generation operations. FERC Order 2000, issued in December 1999, requires utilities to submit plans for transferring operational control of transmission to regional transmission organizations (RTOs) or provide a justification for not doing so. In July 2001, FERC, in an effort to foster large competitive electricity markets, announced a goal of promoting the development of just four RTOs nationwide, including an 11-state RTO in the West that would include California.[3]

Much of the recent debate concerning electricity market restructuring has focused on price spikes, rolling blackouts, and financial difficulties in California and market design and operation to prevent these problems from occurring elsewhere.[4] However, another important consequence of restructuring is the removal of the incentive to maintain sufficient transmission grid capacity that formerly resided with the regulated utilities. Even if the large RTOs envisioned by FERC come to pass, they will need to collect transmission tariffs large enough to generate a sufficient rate of return on investment, as well as to pay for operating and maintenance costs.[5]

HTS Cables and Today's Transmission Grid

The present U.S. transmission system consists almost entirely of overhead lines, which are inexpensive to construct relative to underground cables. The per-circuit-mile cost of underground cables was estimated in 1995 to be approximately $3 million greater than that of overhead lines carrying similar power at similar voltages.[6] HTS cables share this cost disadvantage with

[1]For a discussion of FERC Orders 888, 889, and 2000, see Energy Information Administration, *Electric Power Annual 1999*, Vol. I, Washington, D.C., DOE/EIA-0348(99)/1, August 2000, pp. 23–26.

[2]Energy Information Administration, *The Changing Structure of the Electric Power Industry 2000: An Update*, http://www.eia.doe.gov/cneaf/electricity/chg_stru_update/update2000.html (last accessed March 26, 2002).

[3]Energy Information Administration, *Energy Market Chronology—July 2001*, http://www.eia.doe.gov/emeu/ipsr/chron.html (last accessed March 26, 2002).

[4]See, for example, "Building a Wall Around California," *Energy User News*, Vol. 26, No. 7, July 2001, pp. 18–20, and "Causes and Lessons of the California Electricity Crisis," U.S. Congressional Budget Office, The Congress of the United States, September 2001.

[5]Thompson, A., "The Electric Utility in Transition: Critical Issues and Strategic Responses," IEEE Power Engineering Society Summer Power Meeting, Vancouver, B.C., July 19, 2001.

[6] Fuldner, Arthur H., "Upgrading Transmission Capacity for Wholesale Electric Power Trade," Washington, D.C.: Energy Information Administration, U.S. Department of Energy, Table FE2,

conventional underground cables, so that a prerequisite for HTS cable use where cables are not already underground involves the decision to construct an underground, as opposed to an overhead, transmission circuit. Such a decision might be based upon land use and right-of-way costs, which were not included in the estimate presented earlier, or the desire to bury transmission lines for aesthetic reasons or to reduce their vulnerability to weather, natural disasters, accidents, or sabotage.

For HTS cables, this cost disadvantage might be mitigated in grid regions where no space exists to add additional capacity without increasing voltage. Fuldner (1995) estimates that converting steel tower transmission lines from 115 to 230 kV would cost $500,000 per mile and converting them from 230 kV to 500 kV would cost $800,000 per mile. Converting transmission lines to higher voltage also requires investment in higher-voltage substation equipment such as transformers.

There are approximately 2,000 miles of underground transmission cables in existence today in urban areas. According to the Department of Energy (DOE), about one-quarter of these cables are older than their rated lifetimes.[7] The retrofit of these cables to carry more power, at or near their present voltages, using existing conduits is one target market of the DOE HTS program. Many of these urban areas with existing underground cables also have space and heat constraints resulting from closely sited underground utilities and represent locations in which retrofit with HTS cables would be the only option for expansion without substantial system rebuild or redesign. The installation of a demonstration HTS cable in Detroit Edison's Frisbee substation could be the vanguard of such a program.[8]

In summary, the capability of HTS cables to increase power-transfer capacity without adding new lines or increasing voltage, or to increase transfer capacity at low voltage, is of greatest value in portions of the existing grid for which

1. underground cables already exist and there are space or temperature constraints associated with adding capacity, or

2. transmission constraints exist and the conventional options require new high-voltage transmission lines.

---

http://www.eia.doe.gov/cneaf/pubs_html/feat_trans_capacity/table2.html (last accessed March 26, 2002).

[7]*High Temperature Superconducting Power Products Capture the Attention of Utility Engineers and Planners,* Washington, D.C.: Superconductivity Program for Electric Systems, Office of Power Delivery, U.S. Department of Energy, January 2000.

[8]Chang, Kenneth, "High-Temperature Superconductors Find a Variety of Uses," *New York Times,* May 29, 2001.

This discussion has focused on transmission (defined as voltages above 69 kV), while underground distribution cables (lower than 69 kV) have been a viable option for many communities nationwide for years. However, distribution cables are typically short in length, low in cost, and operated at a variety of utilization levels. These properties mitigate against use of HTS cables. Because they require power to cool them to operating temperature, short HTS cables are not cost-effective because the cooling demands are dominated by the heat loss at the terminations at which they are connected to the ambient-temperature portion of the system. Similarly, HTS cables that are operated at low utilization levels do not transmit enough power to overcome the energy penalty associated with operating their cooling systems. The principal benefit from HTS cables of increased power transfer is typically not an issue for distribution cables. And the cost of cryogenic cooling per unit of power delivered is relatively high for cables carrying only a small amount of power.[9]

Increased Utilization of the Grid

Since the 1970s, new transmission line miles have grown at less than half the rate of electricity demand.[10] Local opposition to new power lines has grown over the years as the country has become more populated and movements have sprung up to protect the aesthetic features of rural areas ranging from pristine forests, wetlands, and prairies to well-settled farming regions. Some communities have come to fear the supposed negative health effects from electromagnetic fields that are present near high-voltage lines.[11] In addition, regulatory uncertainty has inhibited construction of new power lines. It is unclear who will own and operate the transmission system, and how grid improvements will be paid for. At present, rates of return appear to favor construction of new local generation over new transmission connecting to more-distant generation assets.

Transmission line utilization has increased substantially due to the marketing of power from merchant (non-utility-owned) power plants. Many NERC regions

---

[9]Sheahen, Thomas P., Benjamin W. McConnell, and Joseph W. Mulholland, "Method for Estimating Future Markets for High Temperature Superconducting Power Devices," paper presented at the Applied Superconductivity Conference, September 2000, http://www.ornl.gov/HTSC/pdf/ASC00FormatedLongRev.pdf (last accessed April 1, 2002).

[10]Energy Information Administration, *The Changing Structure of the Electric Power Industry 2000: An Update*, http://www.eia.doe.gov/cneaf/electricity/chg_stru_update/update2000.html (last accessed March 26, 2002).

[11]Electromagnetic fields associated with transmission lines have not been conclusively linked to any adverse effects on human health, and no mechanism for the occurrence of such effects at the field levels associated with power lines has been demonstrated. See the National Institute of Environmental Health Sciences, National Institutes of Health fact sheet *Questions and Answers about Electric and Magnetic Fields Associated with the Use of Electric Power,* http://www.niehs.nih.gov/oc/factsheets/emf/emf.htm (last accessed March 26, 2002).

12

have reported a dramatic increase in power-marketing activity and some have reported that such transactions have the potential to cause transmission congestion.[12] According to two different studies, electric-power marketing sales rose from close to zero in 1994 to between 1.0 and 2.5 billion megawatt hours in 1999.[13]

Increased utilization of the grid poses problems for power reliability. Most of the nation's power grid was designed mainly for radial power flow from generation assets to load centers. Growth in power transactions, often over long distances between NERC regions, complicates planning because such transactions are difficult to predict and their effects can be difficult to manage. Such transactions can switch the direction of energy flows across power lines, sometimes resulting in flows that change directions daily or hourly. As has been illustrated by Overbye,[14] a single bulk-power transaction between utilities in different NERC regions can affect the utilization of dozens of high-voltage power lines over a broad geographical area, including lines that do not appear to offer a path between the utilities engaging in the transaction.

Increased utilization of the nation's power grid has led to grid components operating increasingly closer to their technical limits. Line ratings have changed in recent years to accommodate increased power transactions, such that lines can be utilized at full capacity almost year-round.[15] As power flows increase, transmission lines heat and sag toward the ground. Sagging lines can cause outages and fires, and they constitute a safety problem.[16] Furthermore, transmission components are being used longer than was originally intended, with many components older than 30 years and a significant portion older than 50 years.[17] As utilities look to reduce maintenance costs and shift their focus to services unrelated to transmission, in many cases they may be inclined to push the system toward failure.[18]

---

[12]North American Electric Reliability Council, *Reliability Assessment 2000–2009*, Princeton, N.J.: North American Electric Reliability Council, October 2000.

[13]Hirst, Eric, "Expanding U.S. Transmission Capacity," Oak Ridge, Tenn., August 2000, http://www.eei.org/future/reliability/hirst2.pdf (last accessed April 1, 2002). Gale, Roger, Joe Graves, and John Clapp, *The Future of Electric Transmission in the United States*, Washington, D.C.: PA Consulting Group, January 2001.

[14]Overbye, Thomas J., "Reengineering the Electric Grid," American Scientist, Vol. 88, May–June 2000, p. 223.

[15]Hazan, Earl, "America's Aging Transmission System," *Transmission and Distribution World*, May 2000, p. 37.

[16]Seppa, Tapani, "Opportunities and Risks of Higher Utilization of the Transmission System," presentation at Transmission and Distribution World Expo 2000, Cincinnati, April 28, 2000.

[17]Hazan (2000), p. 37.

[18]Hazan (2000), p. 37.

Transmission and Efficient Power Markets

Transmission lines enable the transactions that constitute the electric-power marketplace. A microeconomic study of California power markets conducted prior to the power crisis of 2000 to 2001 concluded that transmission limitations between the northern and southern portions of the state could reduce price competition in the state overall. Significantly, the study found that "relatively small investments in transmission may yield surprisingly large payoffs in terms of increased competition," and "there may be no relationship between the effect of a transmission line in spurring competition and the actual electricity that flows on the line in question."[19] This finding is significant in light of the manner in which transmission planning is conducted, namely that contingency planning may not be attuned to the public's need for power that is both reliable and available at reasonable cost.

A detailed study of power flows in the Northeast United States concluded that transmission limitations would result in substantial price differences among regions.[20] The authors wrote, "Potential for increased trade, while significant, is limited by congestion on the transmission network, especially with regard to the New York-New England interface."

A third study found that today's merchant generation capacity (e.g., generators not owned by the local utility) is typically being connected to existing 115 kV transmission lines, lines that were operated at low utilization in the past and previously were used as "lifelines" during contingencies. Such practices effectively provide a subsidy to new generators who do not have to invest in overall transmission upgrades.[21] The economic trade-off here is that increased use of these low-voltage transmission lines, while increasing their profitability, is degrading transmission reserve margins and hence reliability.

The central impact of transmission constraints is suboptimal dispatch of power generation. When economic generation to meet load requirements creates power flows that would overload lines, grid security coordinators are forced to order

---

[19] Borenstein, Severin, James Bushnell, and Steven Stoft, "The Competitive Effects of Transmission Capacity in a Deregulated Electricity Industry," Stanford, California: Stanford Graduate School of Business, October 1998.

[20] Eynon, Robert T., Thomas J. Leckey, and Douglas R. Hale, "The Electric Transmission Network: A Multi-Region Analysis," *Issues in Midterm Analysis and Forecasting*, Washington, D.C.: Energy Information Administration, U.S. Department of Energy, 2000.

[21] Seppa, Tapani O., "Physical Limitations Affecting Long Distance Energy Sales," presentation at IEEE Power Engineering Society Summer Power Meeting, Seattle, July 18, 2000.

14

curtailment of some generation in order to reduce congestion on power lines.[22] As Hirst, Kirby, and Hadley write, "Because there is no time to take corrective action to prevent cascading failures, it is necessary to preemptively modify the generation dispatch. It is this off-economic dispatch that results in locational price differences."[23]

National policymakers have identified the dearth of transmission investments as a serious concern. As noted previously, FERC has mandated creation of RTOs that would coordinate regional grid planning and possibly assume responsibility for grid investments and grid reliability.[24] Depending upon how RTOs evolve, they could resolve some of the uncertainty surrounding transmission ownership and regulation and could also promote more-aggressive transmission planning and siting.

The topology of the transmission system in relation to generation and load is also important to RTO design. Regions with large amounts of transmission effectively shorten the distance between contiguous areas.[25] Recent studies measure the average distance generators are able to send power by dividing transmission capacity in megawatt (MW) miles by peak power demand. For example, Seppa finds that even as bulk power transactions have increased, the distance generators can send power on average dropped 17 percent from 1984 to 1998, implying that the trade area for a given generator has fallen 30 percent.[26]

Limitations in fuel supplies for power generation may also lead to increased awareness of the need for transmission. While new local generation can be a substitute for increased transmission of power from outside areas, generators are themselves dependent upon a supply of fuel; in the case of most new plants being constructed today, the fuel is natural gas. Transmission and distribution of natural gas has its own constraints, as amply demonstrated by the power crisis in the Western states during 2000 to 2001, during which time a rise in California

---

[22]See Mountford, John D., and Ricardo R. Austria, "Turning Out the Lights," *IEEE Spectrum,* June 1999, p. 36, and Overbye, Thomas J., "Reengineering the Electric Grid," American Scientist, Vol. 88, May–June 2000, p. 224.

[23]Hirst, Eric, Brendan Kirby, and Stan Hadley, "Generation and Transmission Adequacy in a Restructuring U.S. Electricity Industry," paper prepared for Edison Electric Institute, June 1999.

[24]For example, Massey, William L., "FERC and State Roles in Implementing a National Energy Strategy," speech to the National Association of Regulatory Utility Commissioners, Washington, D.C., February 27, 2001.

[25]Reppen, N. Dag, "Impact of RTO Size on the Reliable and Secure Operation of the Eastern Interconnection," paper presented at the IEEE Power Engineering Society Summer Power Meeting, Seattle, July 2000.

[26]Seppa, Tapani O., "Physical Limitations Affecting Long Distance Energy Sales," presentation at IEEE Power Engineering Society Summer Power Meeting, Seattle, July 18, 2000.

natural-gas spot prices (in some cases seven times the average national spot price) caused a concomitant increase in the price of electric power.[27]

Major U.S Transmission Constraints

The decision to install new transmission is based on highly complex data including load-growth models, power-flow models, expected new generation, planned or potential upgrades to other parts of the power grid, expected operating conditions and procedures, and regional and national reliability criteria. Transmission is also installed for many purposes: connecting new generation and growing load centers, reducing congestion for economic reasons, maintaining system reliability, and improving system stability.

Because of the complexity of the system, it is beyond the scope of this study to determine the proper locations for new transmission nationwide today, let alone in 10 or 15 years when HTS cables might provide a substitute for conventional transmission. It should be sufficient to analyze a handful of the most pressing transmission needs today, as identified by NERC and others as having reliability implications. In general, transmission projects that are important enough to be cited in regional or national reports by organizations such as FERC and NERC have been long delayed due to concerns of local residents and officials.

As noted earlier, the average distance served by generators fell 17 percent from 1984 to 1998. Using data available through NERC and EIA, Hirst calculated transmission capacity changes by region from 1998 through 2008.[28]

Overall, NERC data show that the average distance of transmission capacity per unit load will decline a further 12.4 percent by 2008. Of course, actual construction of capacity will differ from this projection. Table 2.1 summarizes Hirst's findings.

---

[27]Sheffrin, Anjali, "Market Analysis Report," public memorandum to the Market Issues/ADR Committee of the California Independent System Operator, January 16, 2001.

[28]Hirst, Eric, "Expanding U.S. Transmission Capacity," August 2000, http://www.eei.org/future/reliability/hirst2.pdf (last accessed April 1, 2002).

16

Table 2.1

Decline in Transmission Capacity by Region

NERC Region[a]	Area Served	Transmission Capacity (MW-miles/MW peak demand)		Percentage Change	
		1998	2008	1989–1998	1998–2008
ECAR	Midwest	201	176	–19.1	–12.7
ERCOT	Texas	116	108	–24.5	–6.4
FRCC	Florida	117	107	–17.9	–8.4
MAAC	Mid-Atlantic	110	94	–11.7	–14.6
MAIN	Illinois and Wisconsin	107	96	–15.7	–10.0
MAPP	Northern Plains	320	265	–14.0	–17.1
NPCC	Northeast	116	101	–5.2	–12.6
SERC	Southeast	172	140	–14.6	–18.8
SPP	Kansas and Oklahoma	134	128	–33.2	–4.5
WSCC	West	411	377	–15.8	–8.2
Total		202	177	–16.2	–12.4

SOURCE: Hirst, Eric, "Expanding U.S. Transmission Capacity," August 2000.

[a]See the Acronyms list for the full names of the regions. A map of U.S. NERC regions may be found at http://www.nerc.com.

If the projections shown in the table hold true, the Southeast region will experience the largest decline in capacity, while the Kansas-Oklahoma region will experience the smallest decline. The Pennsylvania-New Jersey-Maryland (PJM) Interconnection served by the Mid-Atlantic Area Council (MAAC) will have the least transmission capacity per unit peak load, whereas the Western Systems Coordinating Council (WSCC) will have the greatest. The need for transmission is a function of the geographical dispersion of loads and generation; hence, it is natural that the regions with larger areas and more-dispersed population centers have more transmission capacity. Such regional figures are useful only insofar as they may be a rough indication of whether regional transmission systems are keeping pace with the need for transmission in their geographical areas.

The boundaries between planning regions (e.g. between NERC regions or between control areas of formerly vertically integrated utilities) can represent transmission bottlenecks. Although NERC regions coordinate transmission planning activities, the historical development of the system as a radial network may be insufficient to support growing power trade among and across regions. Problems at boundaries may arise because the demand for power transfers in a competitive market may greatly outpace the demand for power transfers that was anticipated between more- or less-self-sufficient entities who exchanged power during emergencies or to rectify temporary imbalances between generation and load within control areas. One way of analyzing regional boundaries is to examine NERC data on transmission-loading relief (TLR) measures, which are instances in which generation is curtailed to reduce congestion along constrained transmission paths. In 2000, FERC undertook an analysis of regional bulk power markets that examined instances of TLR using public logs of these events maintained by NERC.

It is important to note that differences in operating procedures may make this measure a poor one for determining the overall extent of transmission limitations. According to FERC, the lack of uniform standards for implementing TLR measures is a problem for market participants: Some participants may use TLRs to the disadvantage of other generators, TLRs may benefit transmission providers and their affiliates to the detriment of generators, and TLRs increase uncertainty for generators as far as whether their transactions will be completed as planned.[29] TLR events can often be mitigated by upgrades to substation and other equipment. However, these events do give some indication of localities where constraints exist.

Specific TLR measures were discussed in FERC's 2000 studies of the Midwest and Southeast United States bulk-power markets. The Midwest experienced 492 TLR events in 2000, up from 86 in 1999. The region identified ten flowgates where TLRs were most frequent. The most significant TLR events took place around the West Virginia-Virginia region where American Electric Power (AEP) has proposed a 765 kV line to alleviate transmission constraints and along a variety of 345 kV lines to the south, west, and southeast of St. Louis, Missouri.

The Southeast region also experienced a dramatic upsurge in the number of TLR measures in 2000, with 184 events in that year.[30] Weather and increased power imports contributed to congestion. FERC lists ten flowgates where major TLRs were called. The three most significant of these were the Volunteer-Phipps Bend 500 kV line owned by the Tennessee Valley Authority (TVA), which connects eastern Tennessee to southwest Virginia, the Webre-Richard line connecting southwest Louisiana to New Orleans, and the Hot Springs-McNeil 500 kV line

---

[29]Federal Energy Regulatory Commission, *Investigation of Bulk Power Markets: Midwest Region,* Washington, D.C.: Federal Energy Regulatory Commission, November 1, 2000, pp. 2–35.

[30]Federal Energy Regulatory Commission, *Investigation of Bulk Power Markets: Southeast Region,* Washington, D.C.: Federal Energy Regulatory Commission, November 1, 2000, pp. 3–16.

connecting central Arkansas to southern Arkansas. TLR log files confirm that these TLR events constitute the bulk of TLR-logged events nationwide.[31]

FERC's 2000 bulk power studies of the West and Northeast discussed transmission limitations in terms of constrained paths or regions. In the case of the Northeast, FERC identified the path into New York City and Long Island as seriously constrained with implications for reliability during the next few years.[32] New York City is fed by outside lines running in from the north along the Hudson River and from the west through New Jersey. The city's load can exceed 10 gigawatts (GW) and the city's generation totals just under 8 GW. Transmission of power into New York City is 5,175 MW, but those lines are vulnerable to thunderstorms. Transmission into Long Island is 975 MW, with another 330 MW scheduled to come on-line when the Cross-Sound Cable is completed. Transmission into Long Island is also vulnerable to thunderstorms and is constrained by the dispatch of generation located in New York City.

In the West, the primary transmission constraints identified by FERC were discussed in terms of their contribution to the California crisis. In 2000 and 2001, transmission into California was reduced due to higher demand in other western states; thus, the main congested lines were Path 15 (the northern portion of the link between northern and southern California) and Path 26 (the southern portion of the north-south link).[33] Path 15 has been identified as a major regional constraint in the National Energy Policy Report, which recommended authorizing "the Western Area Power Administration to explore relieving the 'Path 15' bottleneck through transmission expansions financed by nonfederal contributions."[34]

Path 15 is composed of two 500 kV and four 230 kV overhead transmission lines between Midway and Los Banos, California. The Secretary of Energy has recently announced a $300 million Path 15 upgrade project involving Pacific Gas and Electric Corporation (PG&E), an investor-owned utility that owns the Path 15 lines and has responsibility for planning and investing in line upgrades; the

---

[31]Raw TLR log files can be obtained from NERC at http://www.nerc.com/~filez/logs.html. Charts available at the site indicate that most TLR events take place in the AEP, MAIN, SPP, and TVA security coordination areas, in that order.

[32]Federal Energy Regulatory Commission, *Investigation of Bulk Power Markets: Northeast Region*, Washington, D.C.: Federal Energy Regulatory Commission, November 1, 2000, pp. 1–72.

[33]Federal Energy Regulatory Commission, *Investigation of Bulk Power Markets: Part I*, Washington, D.C.: Federal Energy Regulatory Commission, November 1, 2000, pp. 3–29.

[34]National Energy Policy Development Group, *Reliable, Affordable, and Environmentally Sound Energy for America's Future*, Washington, D.C.: U.S. Government Printing Office, May 2001.

Transmission Agency of Northern California; and six other organizations.[35] The Path 15 lines, which roughly connect northern and southern California, have a power-transfer capacity of approximately 3,600 MW, and have historically experienced highest flows during the winter season, with lower flows during the spring. The planned upgrade will increase Path 15 capacity by 1,500 MW.

Regional transmission customers identified Path 15 as a constraint at which loading relief measures have been necessary, with 39 such events experienced in 1999.[36] California operates a congestion-pricing system in which transmission customers bid on the right to schedule flows over Path 15 during times of high congestion. In 2000 and 2001, California looped power around the congested Path 15 using the DC Intertie connecting Southern California to the Northwest, bringing power back down into Northern California using the AC Pacific Intertie. This procedure contributed to blackouts in California on one occasion, January 21, 2001, when a computer crash brought down a converter station in Oregon leading to a 20-minute outage of 120 MW.[37] On April 1, 2001, wind knocked down six towers on the DC Pacific Intertie leading to a Stage 2 emergency declaration by the California Independent System Operator (ISO).[38]

The California ISO identified Path 15 as one of ten transmission projects required in California in the coming years. In a memorandum to the California legislature, the ISO recommended a new 500 kV loop from Midway to Los Banos through Gregg at a cost of $250-$350 million.[39] The ISO suggested that the line could be operational by 2004 with expedited permitting and planning. The ISO has also supported PG&E Path 15 upgrade proposals that include new 500 kV lines between Los Banos, Gates, and Midway.[40]

NERC's October 2000 national reliability report identified two specific major transmission projects of regional importance: the 765 kV West Virginia to Virginia transmission line proposed by AEP and a Minnesota to Wisconsin

[35] "Energy Secretary Abraham Announces $300 Million Deal to Upgrade California's Path 15," DOE press release, October 18, 2001, http://www.energy.gov/HQPress/releases01/octpr/pr01180.htm (last accessed March 26, 2002).

[36] Northwest Regional Transmission Association, Southwest Regional Transmission Association, and Western Regional Transmission Association, *Western Interconnection Biennial Transmission Plan,* July 2000.

[37] Stipp, David, "The Real Threat to America's Power," *Fortune,* March 5, 2001, p. 138.

[38] "U.S. West Power Line Shut, California Calls Power Alert," Reuters, April 2, 2001.

[39] Winter, Terry M., "Memorandum to Senate Committee on Energy, Utilities & Communications," Assembly Utilities and Commerce Committee, August 10, 2000.

[40] "Testimony of Armando J. Perez, Stephen Thomas Greenleaf, and Keith Casey on Behalf of the California Independent System Operator Before the Public Utilities Commission of the State of California," Application 01-04-012, September 25, 2001.

transmission line.[41] In addition, two other NERC regional reports discuss specific transmission constraints that could require construction of more than one line: the northeast New Mexico area and the connection from south to north Texas where transmission loading relief will be necessary to alleviate congestion in the near term.[42]

A 2000 study employing telephone interviews with utility managers to identify transmission facilities that were not being built identified the following projects as among the most important: (1) the 765 kV West Virginia-Virginia line; (2) the 230 and 345 kV lines proposed to connect Minnesota and Wisconsin; (3) a 500 kV line between Georgia and Florida; (4) expanded capacity between Michigan and Indiana; (5) expanded interfaces between the PJM, New York, and New England areas; (6) transmission between Wyoming and eastern Colorado; and (7) load pockets in and around Boston, New York City, Long Island, San Francisco, and San Diego.[43]

The DOE's Power Outage Study Team (POST) has also conducted research on outages that relates to transmission and distribution constraints.[44] Most of the outages studied by POST did not result from transmission constraints; however, some transmission and distribution limitations were discussed in the report in conjunction with outages that were related to plant shutdowns during times of high load. In Chicago on July 30 through August 12, 1999, underground distribution cables overloaded during a period of high load: "Two separate cable faults at the Northwest substation de-energized transformers, thereby overloading nearby interconnected transformers and causing them to shut down." According to POST, a July 9, 1999, outage was related to the loss of eight feeder cables in the Washington Heights network: "The loss of feeders occurred because of heat-related failures in connections, cables, and transformers." Voluntary curtailments and loss of service in Long Island took place between July 3 and July 8, 1999, due to hot weather and load growth on the South Fork of Long Island, which "was on the edge of voltage collapse." In the Delmarva (Delaware, Maryland, and Virginia) Peninsula on July 6, 1999, plant outages and high loads "produced a capacity shortfall that could not be remedied through

[41]North American Electric Reliability Council, *Reliability Assessment, 2000–2009*, Washington, D.C.: NERC, October 2000.

[42]North American Electric Reliability Council, October 2000.

[43]Hirst, Eric, "Expanding U.S. Transmission Capacity," August 2000, http://www.eei.org/future/reliability/hirst2.pdf (last accessed April 1,2002).

[44]Power Outage Study Team, *Report of the U.S. Department of Energy's Power Outage Study Team: Findings and Recommendations to Enhance Reliability from the Summer of 1999*, Washington, D.C.: U.S. Department of Energy, March 2000.

energy imports on the transmission system." Some of these contingencies may point to areas where more transmission capability is needed.

Figure 2.2 indicates schematically the areas of the United States in which major transmission constraints have been identified.

Technology Alternatives: High-Voltage Direct Current and Flexible AC Transmission System Devices

HTS cables will be likely candidates to address the transmission constraints indicated in Figure 2.2 in areas where underground cables are an option and in areas where HTS cables provide benefits that are not possible with conventional cables.

As noted previously, HTS cables are likely to be particularly useful in areas with space or temperature restrictions, such as urban areas, but they may also be feasible in rural locations where new overhead lines face difficult siting challenges and there is a need for long, high-capacity, high-utilization cables. An important benefit in this case, which is quantified in Chapter 3 by the computer

RAND*MR1531-2.2*

NOTE: Major transmission bottlenecks as identified in NERC reports, TLR log files, POST studies, press accounts, and national energy reports.

Figure 2.2—Areas of the United States in Which Major Transmission Constraints Have Been Identified

simulation results for California's Path 15, is to provide a parallel transmission path at a lower voltage to increase power-transfer capacity. We note, however, that the HTS cables required in this case would not be available in the near term because they are much longer than current demonstrations and would require the development and demonstration of cooling stations at periodic intervals, as well as quantities of HTS wire beyond current manufacturing capabilities.

There are also other technologies that may resolve these transmission constraints. Conventional approaches include upgrading conductors on overhead lines or addressing impedance or reactance problems with capacitors or inductors.[45] Flexible AC transmission system (FACTS) devices represent an evolving technology that enables transmission operators to direct power flows across individual lines. FACTS devices use semiconductor switches to dynamically adjust transmission line parameters in order to control flows and help alleviate congestion, reduce power quality problems, and generally aid in the economically efficient dispatch of generation. A FACTS device is currently being demonstrated in upstate New York routing power from Quebec to New York City.[46] FACTS devices may be used in combination with higher-capacity HTS cables, but because FACTS can help postpone grid improvements they may also represent a competing technology.

High-voltage direct current (HVDC) cables are another potential technological alternative to HTS cables. For example, the Connecticut Siting Council recently approved the TransÉnergie Cross-Sound Cable Project, a 24-mile cable system from New Haven, Connecticut, to Brookhaven, New York, that will carry up to 330 MW of DC power between Connecticut and Long Island beneath the seabed of the New Haven Harbor Federal Navigation Channel and Long Island Sound.[47] The Cross-Sound Cable was expected to begin operation in the summer of 2002. However, Connecticut government officials are considering a moratorium on construction.[48] Another proposed large regional transmission project—the Neptune Project—to connect New Brunswick and Nova Scotia to major urban load centers in Boston, Long Island, Connecticut, and New York also envisions

---

[45]Fuldner, Arthur H., "Upgrading Transmission Capacity for Wholesale Electric Power Trade," Washington, D.C.: Energy Information Administration, U.S. Department of Energy, Table FE2, http://www.eia.doe.gov/cneaf/pubs_html/feat_trans_capacity/table2.html (last accessed March 26, 2002).

[46]Fairley, Peter, "A Smarter Power Grid," *Technology Review,* July/August 2001, pp. 41–49.

[47]"Connecticut Siting Council Approves Cross-Sound Cable Project," press release, January 3, 2002, http://www.crosssoundcable.com/NewFiles/pressrel.htm (last accessed January 22, 2002).

[48]Caffrey, Andrew, "Northeast States in Power Play: Connecticut Looks to Block Underwater Electrical Line to New York's Long Island," *Wall Street Journal,* April 17, 2002, p. B10

the use of undersea HVDC cables.[49] The first phase of the Neptune Project is envisioned as a short connection from New Jersey to New York City and Long Island with anticipated completion in summer 2003.

HVDC cable technology is being developed and implemented by both ABB (formerly Asea Brown Boveri) and Siemens. The ABB product, known as HVDC Light, uses DC cables with extruded polymer insulation, high-speed semiconductor switches and converters, and low-cost installation by plowing instead of conventional trench excavation. It is being implemented in several underground and submarine applications throughout the world.[50] The low-cost installation and AC-DC conversion advances associated with HVDC technologies may also support the development of HTS technologies by lowering the cost of underground cable installation and facilitating the use of DC transmission. HTS DC cables have zero conduction loss, although they will nevertheless have thermal losses that will require a cryogenic cooling system for heat removal.

---

[49]"Neptune Regional Transmission Project Plans Huge Electricity Transmission Grid," *Natural Gas & Energy News,* July 2001.

[50]Asplund, Gunnar, Kjell Eriksson, and Ove Tollerz, "Land and Sea Cable Interconnections with HVDC Light," paper presented at CEPSI 2000 conference in Manila, Philippines, October 23–27, 2000.

3. Potential Impact of HTS Cables on Grid Reliability and Available Transfer Capacity

This chapter describes the results of computer-generated simulations of power flows in two specific locations of the U.S. electrical transmission grid, with and without the presence of HTS cables. The objective of these simulations was to quantify the impact of HTS cable use in real-world applications. We chose these two cases to investigate because they represent two very different types of applications, both of which address an identified U.S. transmission constraint.

The first case represents a heavily loaded urban network, the downtown Chicago area, in which space constraints may preclude conventional transmission growth options. The additional power-carrying capability of HTS cables may provide a means to serve load growth by replacing, and hence upgrading, conventional underground cables at 138 kV using existing underground conduits. The use of the existing conduits would preclude the need for the excavation and construction associated with adding conventional cables to increase power-transfer capacity.

The second case represents a constrained transmission path , Path 15 in California, between major load centers. Path 15 is the major route through which power is transferred between northern and southern California. In this case, HTS cables could provide a means to increase transfer capacity without construction of additional 500 kV transmission lines in an already highly loaded corridor. Because of their higher power-carrying capacity, 230 kV HTS cables could be used to form a parallel path instead. The parallel path improves reliability and the lower voltage avoids additional expenses from high-voltage substation equipment. However, unlike the Chicago case, this case does not represent a near-term application of HTS cables because it requires cable lengths that are much longer than those currently being demonstrated.

The computer simulations reported here were performed for RAND under subcontract by PowerWorld Corporation using a model called PowerWorld Simulator,[1] which provides a visualization of power flows in a transmission grid

---

[1]A detailed description of PowerWorld Simulator and examples of its use are available at http://www.powerworld.com. See also, Overbye, T. J., and J. D. Weber, "New Methods for the

26

based upon input on the generators, loads, transmission lines, transformers, and other electrical power equipment connected to the grid. To input the required data, PowerWorld Simulator has the ability to read publicly available power-flow cases such as those formerly published by FERC from the Form 715 filings of the NERC regions.[2]

The calculation of flows in the grid is based on Kirchhoff's Laws. The analysis of the circuit centers on the buses, or nodes, where generation, transmission, and loads are connected. The net flow of power at each bus must be identically zero to ensure conservation of energy. Based upon the input data, the loads, the generation, or some combination of the two is fixed. For the entire circuit, generation must then equal demand from the loads. (This need not be true locally, because conductors can carry excess generated power to areas where demand exceeds supply.) By calculating how power flows to satisfy Kirchhoff's Laws, one obtains information about the power flows through the conductors that make up the grid.[3] PowerWorld Simulator provides a variety of visualizations of these power flows, including indications of magnitude, direction, and percentage of available current and real and reactive power[4] set by input limitations on components.

Simulations of the Downtown Chicago Transmission Grid

The input data for the Chicago simulations described here were obtained from the Mid-America Interconnected Network (MAIN) Form 715 submission to the Federal Energy Regulatory Commission[5] for the summer 2000 peak load. In order to reduce the size of the simulation while still maintaining the important features of the interconnected network, the portion of MAIN outside of the

Visualization of Electric Power Systems Information," July 2000, http://www.pserc.wisc.edu (last accessed April 1, 2002).

[2]FERC no longer publishes the Form 715 data for national security reasons, but provides information on how to obtain the data through a Freedom of Information Act (FOIA) request.

[3]For a detailed description of power-flow solution methods, see Wood, Allen J., and Bruce F. Wollenberg, *Power Generation Operation and Control*, 2nd Ed. John Wiley & Sons, New York, 1996, Chapter 4.

[4]In an AC circuit, current leads or lags the voltage. The current component that is in phase with the voltage transmits real power and the out-of-phase component is associated with reactive power, which represents energy stored in electric or magnetic fields. Reactive power is consumed or absorbed in the magnetic field of inductive equipment. This consumption tends to depress transmission voltage, which can create voltage control problems. Reactive power can be transmitted only over short distances. If reactive power cannot be supplied promptly in an area of decaying voltage, voltage collapse can occur. See Taylor, Carson W., "Improving Grid Behavior," *IEEE Spectrum*, June 1999, pp. 40–45.

[5]See http://www.ferc.gov/electric/f715/data/form715.htm for information on how to file a FOIA request for these data.

27

control area of Commonwealth Edison, the local Chicago utility company, was treated as an equivalent circuit. This reduced the size of the simulation from 30,000 buses to 1,000 buses. The total load for the simulation was 19,000 MW and 7,000 MVAR (megavolts amperes reactive)[6], and the simulated network included 138 kV and 345 kV transmission lines.

To determine the impact of HTS cables on reliability, we focused the analysis on two smaller regions: the *Area of Interest,* a region of downtown Chicago in which transmission cables are currently operated at close to their rated capacity and may become overloaded in the event of contingencies such as transmission line outages, and the *Contingency Area,* within which we define a list of transmission lines that will be taken out of service in contingency simulations.[7] Figure 3.1 shows a portion of the simulated region, indicating both the Area of Interest and

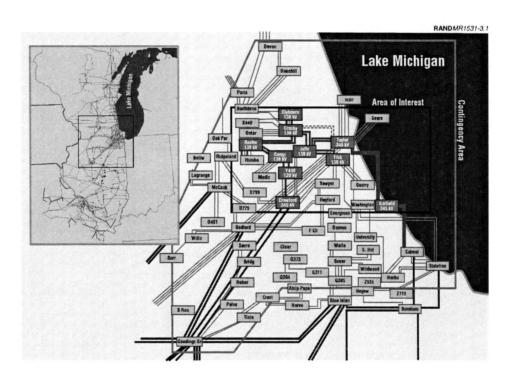

Figure 3.1—Portion of Downtown Chicago Grid Showing Area of Interest and Contingency Area, Commonwealth Edison Control Area Within MAIN Transmission Grid Shown in Inset

---

[6]Real power is measured in MW and reactive power is measured in MVAR.

[7]Weber, J. D., "Efficient Available Transfer Capability Using Linear Methods," May 2001, available at http://www.pserc.wisc.edu (last accessed April 1, 2002).

the Contingency Area.[8] The inset in the figure shows the grid within the entire MAIN region. The box in the inset surrounds the Commonwealth Edison control area, and within that box is the Contingency Area.

The shaded boxes in Figure 3.1 represent substations, which may contain one or more buses. The lighter lines represent transmission lines at 138 kV and the darker lines represent transmission lines at 345 kV. The darker shaded substations represent those with significant loads and components of important transmission paths. The maximum voltage bus is indicated for each of these substations. The darker transmission lines within the Area of Interest are 138 kV underground cables (and two 345 kV underground cables) that were replaced in some of the simulations by HTS cables, as described in the following.

We performed N-1 contingency analyses by taking out of service each of the 133 transmission lines in the Contingency Area, one at a time, running the simulation model to obtain the power flow, and monitoring all elements in the system for violations, which were defined as loadings of more than 100 percent of the MVA rating[9] for transmission lines and transformers or voltages outside the range 90-110 percent of nominal value (0.9 to 1.1 per unit, or p.u.) for a bus. We then recorded the number of violations that occurred for each contingency, which allowed the identification of the transmission lines that had the largest number of violations and the highest overload for all of the contingencies. These lines are candidates for replacement with HTS cables with higher MVA ratings.

Table 3.1 lists the transmission lines (all underground cables) that had the largest overloads and consistently showed overloads under different N-1 contingency simulations. The table lists the substations (see Figure 3.1) from which and to which each of these underground cables transmits power, as well as their length, voltage, and nominal current rating in amperes and power rating in MVA. The differences in ratings may reflect differences in cable design, location, and configuration and are typically based upon temperature limitations that arise from heating of the conductors due to electrical losses.

---

[8]The description of the contingency analysis follows Ettedgui, E., S. Grijalva, R. Silberglitt, and C. Wakefield, "Potential Impact of High Temperature Superconducting Cables on Grid Reliability and Available Transfer Capacity: Power Flow Simulation Results," Institute of Electrical and Electronics Engineers (IEEE) Power Engineering Society Winter Meeting, January 2002, available on CD from the IEEE.

[9]The rating in MVA is the maximum value of the power flowing in the three-phase AC circuit, which is given by volts•amperes• $\sqrt{3}$.

Table 3.1

Transmission Lines Identified in N-1 Contingency Analysis

Substation		Length	Voltage	Rating	Rating
From	To	(Miles)	(Volts)	(Amperes)	(MVA)
Crosby	Jefferson	1.66	138,000	920	220
Garfield	Taylor	4.47	345,000	1,171	700
Calumet	Garfield	3.07	345,000	2,276	1,360
Fisk	Jefferson	2.23	138,000	1,360	325
Jefferson	Taylor	2.09	138,000	837	200
Crosby	Rockwell	1.61	138,000	920	220
Charter	Congress	0.49	138,000	1,326	317
Clybourn	Crosby	0.89	138,000	251	60
Congress	Rockwell	0.59	138,000	1326	317
Humbolt	Rockwell	1.47	138,000	251	60

We note that the overloads under N-1 contingencies in these simulations do not
consider remedial actions that grid operators may take during contingencies,
including, for example, adjusting phase shifters, opening or closing breakers,
changing the local generation profile, and employing demand-side management
schemes. We were not able to include such actions because information
concerning operating guidelines under contingencies tends to be confidential.
Nevertheless, our results identify the impact that replacing specific transmission
lines with HTS cables in a real utility system would have on reliability and
transmission capacity, irrespective of specific operating procedures that may also
be used in the event of contingencies. The most serious violations, for example
those with overloads greater than 130 percent, may represent cases in which
operating measures would be insufficient to mitigate overloads, so that the
transmission lines in question may require uprating, and replacement with HTS
cables could serve as an alternative measure that avoids having to build new
conduits.

Transmission lines are typically modeled in power-flow simulations using a pi-
model for the cables or conductors. In this model, the cable or conductor consists
of a resistor and inductor in series connected to two capacitors in parallel at
either end, so the cable or conductor resembles the Greek letter pi. The
inductance, capacitance, and resistance of each conventional cable or conductor
shown in Figure 3.1 are known input data. In order to incorporate HTS cables in
the simulation, their resistance, inductance, and capacitance, as well as their
maximum current rating, were required. These parameters were provided by
Pirelli Cables and Systems for HTS cables of appropriate voltage and length that
could be substituted in the simulation for the conventional cables listed in Table

30

3.1.[10] The HTS cable resistance was nearly zero and the HTS cable capacitance and inductance were much lower than those of the conventional cables. These small values provided a convenient redistribution of power flows by attracting flow through the HTS cables, and the higher current rating of the HTS cables (2,500 amperes) allowed additional flow through these cables without overloads.

The HTS cable substitutions included three additional changes to the system. First, because the downtown Chicago grid has two almost parallel electrical systems (see Figure 3.1), for consistency all parallel lines were replaced with HTS cables. This added 8 cable replacements to the 11 listed in Table 3.1. (We note that the Clybourn to Crosby entry in Table 3.1 represents two cables because it is a double circuit and both circuits had overloads under N-1 contingencies.) Second, we found it necessary to uprate the range of two transformers in the Crawford substation from 339 MVA and 420 MVA to 500 MVA as a result of the increased power flow through the HTS cables. Finally, one conventional line (Crawford to Y450) was overloaded after the HTS cable substitutions were made, so that we also replaced this with an HTS cable, for a total of 20 HTS cable replacements.

After making the HTS cable replacements (the darker lines within the Area of Interest in Figure 3.1), we repeated the N-1 contingency analysis to determine the impact on reliability. Table 3.2 shows the results for the most severe contingencies. In the table, "Base" refers to the case without the HTS cable replacements and "HTS" refers to the case with the HTS cable replacements. The "Contingency Location" column indicates which transmission line was removed from service under that particular contingency. In some cases, the transmission line was within a substation.

Each row in Table 3.2 represents a distinct contingency. In the simulation model, each contingency is identified by the buses to which and from which the transmission line carries power and the designation and number of its circuit. These labels have been suppressed in Table 3.2. We note that the replacement of conventional transmission lines with HTS cables consistently decreased the number and severity of violations in the system. The base case had a total of 105 violations, whereas the case with the HTS replacements had only 37 violations. If we define the number of violations under N-1 contingency analysis as a reliability metric, then the HTS replacements increased reliability according to this metric by a factor of approximately 3.

---

[10]Nathan Kelley, private communications, Pirelli Cables and Systems, Lexington, S.C. The HTS cable design was the "cold dielectric," in which both conductors and insulation are maintained at cryogenic temperature within a cooled enclosure. See Appendix A for details.

Table 3.2

Contingency Analysis Results for the Most Severe Contingencies

Contingency Location	Number of Violations		Maximum Line Loading (Percentage)	
	Base	HTS	Base	HTS
Taylor (within substation)	5	5	129.1	109.0
Taylor (within substation)	5	5	139.5	108.8
D799-Ridge	5	1	111.4	101.6
Crawford-D799	5	1	113.5	103.9
Crosby-Jefferson	4	3	164.1	102.9
Crawford-Goodings	6	3	153.2	105.8
Fisk-Jefferson	3	1	127.7	100.0
Fisk-Jefferson	3	3	138.5	101.4
Jefferson-Taylor	3	1	113.6	101.0
Crawford (within substation)	6	3	110.6	104.6
Crawford (within substation)	4	1	131.2	107.2
Crawford-Jefferson	3	0	103.8	—
Crawford (within substation)	3	0	103.8	—
Crosby-Jefferson	6	3	163.2	102.8
Crawford-Fisk	3	0	107.2	—
Jefferson-Taylor	3	1	129.8	102.0
Crawford-Fisk	5	2	105.8	101.5
Hayford-Sawyer	3	0	107.8	—
Crawford (within substation)	5	2	105.8	101.5
Fisk-Quarry	2	0	103.0	—
Fisk (within substation)	2	0	131.7	—
Congress-Rockwell	2	0	150.8	—
Crosby-Rockwell	2	0	132.3	—

The results shown in Table 3.2 demonstrate that the replacement of a substantial number (20) of heavily loaded transmission lines in downtown Chicago with HTS cables would garner substantial reliability benefits. What about replacement of a smaller number of lines? To address this question, we note that the only substations that bring power into the Area of Interest (from the south) at 345 kV and transform it to 138 kV are McCook, Crawford, and Taylor, and that the path from McCook to the load substations in the Area of Interest (the darker shaded boxes) also goes through Crawford. Moreover, Table 3.2 shows that the most severe contingencies were those between Crosby and Jefferson that effectively cut the only alternate path through Taylor. This suggests that enhancing the path through Taylor with HTS cable replacements or creating a parallel path with additional HTS cables may provide reliability benefits.

We performed power-flow simulations and N-1 contingency analysis for the following cases: (1) addition of two HTS cables from Taylor to Crosby, as shown

32

by the darker dotted lines in Figure 3.1; and (2) two HTS cable replacements from Jefferson to Crosby.[11] In these simulations, all other transmission lines in the Area of Interest shown as darker in Figure 3.1 remained conventional lines with parameters as obtained from the Form 715 input data. Table 3.3 shows the results for the most severe contingencies, where "Base Case" refers to the system without HTS cables, "Two HTS Additions" refers to case (1) and "Two HTS Replacements" refers to case (2) just noted. As in the 20-HTS-cable case, the addition or replacement of just two HTS cables significantly reduces the number and severity of violations.

The simulation results showed that the number of violations was reduced from 91 in the base case to 43 for two HTS additions or 73 for two HTS replacements, and the average overload was reduced from 118 percent in the base case to 107 percent for two HTS additions or 108 percent for two HTS replacements. We note that the total number of violations in the base case is slightly different than in the base case represented in Table 3.2 because a slight modification was made (opening of one circuit breaker) in the Table 3.3 base case to reduce the loading on one of the McCook to Ridgeland circuits.

Returning to the 20-HTS-cable case, rather than accepting increased reliability, we can use the reduced number of violations with the HTS cable replacements to increase the transfer of power into the Area of Interest up to a level at which the number and severity of violations is similar to that with the conventional cables. We define a load-injection group as the set of buses at which the load will be increased. For this simulation, the load injection group included buses in the Clybourn, Crawford, Jefferson, Sears, and Fisk substations. Additional power transfer was assumed to come from a seller injection group in the Southwest region of the Commonwealth Edison control area. The additional load was brought into the Area of Interest via the 345 kV lines into the Crawford substation.

Table 3.4 shows the results of these simulations, which indicate that 41 percent additional load, corresponding to 365 additional MW, could be served in this area of Chicago in the case with 20 HTS cable replacements (the darker lines within the Area of Interest in Figure 3.1), *while maintaining the same level of reliability as in the base case* (all conventional transmission lines).

---

[11]These simulation results are not described in the Ettedgui et al. (2002) IEEE document referenced earlier. The simulations were performed for RAND by Santiago Grijalva of PowerWorld Corporation.

Table 3.3

Contingency Analysis Results for the Most Severe Contingencies with Two HTS Cable Additions or Replacements

Contingency Location	Base Case		Two HTS Additions		Two HTS Replacements	
	Violations	Maxi-mum %	Violations	Maxi-mum %	Violations	Maxi-mum %
Crawford (within station)	9	130	7	119	8	121
Crosby-Jefferson	8	119	0	—	8	119
McCook-Ridgeland	6	116	5	114	6	115
Crawford-Goodings	6	114	5	108	6	113
Crawford-D799	6	110	5	107	6	109
Crosby-Jefferson	4	117	0	—	4	117
Crawford (within station)	4	110	2	108	4	110
D799-Ridgeland	4	108	3	105	4	107
Fisk-Jefferson	4	104	2	101	4	103
Y450-Congress	3	172	1	116	2	115
Y450-Crawford	3	172	1	116	2	115
Congress-Rockwell	2	156	1	108	2	107
Y450-Congress	1	165	0	—	0	—
Y450-Crawford	1	165	0	—	0	—
Congress-Rockwell	1	142	0	—	0	—
D799-Ridgeland	1	141	0	—	0	—
Crawford-D799	1	138	0	—	0	—
Crosby-Rockwell	1	126	0	—	0	—
X440-Crosby	1	123	0	—	0	—
X440-Northbrook	1	122	0	—	0	—
Crosby-Rockwell	1	116	0	—	0	—

We note that the additional required power would use the higher capacity of the HTS cables that are installed as retrofits using existing conduits, and thus would not require costly excavation to install new underground cables, as discussed previously, and would also avoid the impact of such excavation on local

Table 3.4

Impact of Increased Load on Reliability

Simulation Case	Total Load (MW)	Load Increase (MW)	Load Increase (Percentage)	Number of Violations
Base case	890	0	0	105
20 HTS cables	890	0	0	37
20 HTS cables	1,068	178	20	61
20 HTS cables	1,255	365	41	104

businesses and traffic. We also note that, despite the increased power flowing through the HTS cables, the N-1 contingency analysis, including the HTS cable contingencies, shows that the same level of reliability is maintained. This is because the HTS cables would provide parallel paths to relieve stress on what was formerly a single heavily loaded path.

Simulations of California's Path 15 Transmission Circuit

The input data for the California Path 15 simulations described here were obtained from the WSSC Form 715 submission to FERC[12] for the Winter 2001 peak load. Figure 3.2 shows a portion of the California transmission grid including Path 15. As in Figure 3.1, the boxes represent substations, which may contain one or more buses, transformers, and other power equipment. The darker lines with the arrows below, and above and to the left of, the Gates substation at the center of the figure represent transmission lines at 500 kV. The darker lines with the arrows above and to the right of the Gates substation represent transmission lines at 230 kV. The lighter solid lines with arrows represent 230 kV transmission lines that were replaced with HTS cables in some of the simulations and the lighter dashed lines with arrows represent 230 kV HTS cable additions. The darker shaded substation boxes are connected through Path 15.

Path 15 is composed of two 500 kV transmission lines from Midway to Gates and Los Banos, plus four 230 kV transmission lines, two each from Gates to Panoche and from Gates to Henrietta T. The base case simulation treated all of these as conventional overhead transmission lines, with input data for their pi-models provided in the Form 715 submission. In order to obtain the Path 15 transmission constraint, a simulation was performed in which imports to northern California from the Bonneville Power Administration were reduced and exports from

---

[12] See http://www.ferc.gov/electric/f715/data/form715.htm for information on how to file a FOIA request for these data.

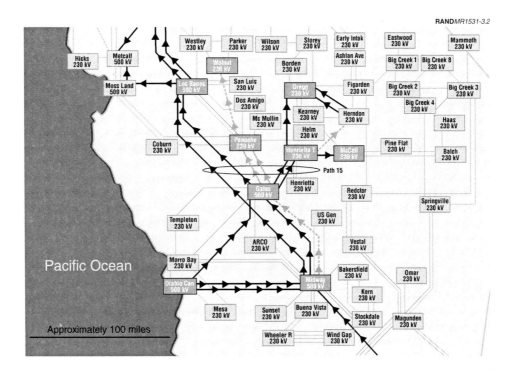

RAND*MR1531-3.2*

Figure 3.2—Portion of the California Grid Showing Path 15

Southern California to Northern California were simultaneously increased, until transfer capacity limits were reached.[13] The maximum flow was 3,740 MW, of which 2,857 MW went through the two 500 kV transmission lines, which were loaded almost to capacity. Table 3.5 shows the most severe contingencies, indicating that the outages of these 500 kV lines are responsible for the greatest number of violations, with maximum overloads up to 175 percent.

HTS cables, with their ability to attract increased power flow and their increased capacity to carry additional power, as described in the previous section, present an opportunity to relieve the stress on the 500 kV portion of Path 15 by providing a parallel 230 kV path. This is possible without transformer upgrades because the 500 kV to 230 kV transformers at the Gates and Midway substations are loaded to only 50 percent and 35 percent of their capacity, respectively, in the base case.

The parallel path using 230 kV HTS cables, shown in Figure 3.2 as the lighter lines with arrows, replaces both conventional 230 kV transmission lines from Gates to Panoche with HTS cables and adds three new 230 kV HTS cables from Midway to Gates, from Panoche to Walnut, and from Henrietta T to Herndon.

---

[13] These simulations were performed for RAND by Santiago Grijalva of PowerWorld Corporation.

Table 3.5

Contingency Analysis Results for the Most Severe Contingencies for Path 15 at Maximum Operating Limit

Contingency	Number of Violations	Maximum Line Loading (Percentage)
Los Banos (within substation at 500 kV)	9	175
Los Banos (within substation at 500 kV)	9	175
Los Banos-Midway (500 kV)	9	175
Henrietta-Gates (230 kV)	7	106
Los Banos-Gates (500 kV)	7	160
Los Banos (within substation at 500 kV)	7	160
Gates-Gates (within substation at 500 kV)	5	123
Gates-Midway (500 kV)	5	123
Storey-Gregg (230 kV)	4	116
Gates-Gates (within substation at 500 kV)	4	116
Gates-Gates (within substation 500 to 230 kV)	4	116
Metcalf-Moss Landing (500 kV)	4	127

The new HTS cables allow transfer of power transformed to 230 kV at Midway through Path 15 and allow power to be sent north of Panoche without further loading of buses in the Los Banos substation, which, for example, would result from any of the proposed 500 kV upgrades involving Los Banos, Gates, and Midway as described in Chapter 2. The replacement HTS cables from Gates to Panoche increase Path 15 capacity to transmit the additional flow from Midway, and the new HTS cable from Henrietta to Herndon increases flow north to Gregg.

The use of these five HTS cables (two replacements and three additions) would increase the transfer capacity through Path 15 to 5,162 MW, an increase of 1,422 MW, or 38 percent, *while maintaining the same level of reliability as in the base case* (122 violations under N-1 contingencies with a maximum line loading of 166 percent). In addition, the parallel 230 kV path would provide a lower vulnerability option than would increasing the capacity of the 500 kV lines because that path would reduce the impact of localized damage to transmission lines or substations from weather, natural disasters, accidents, or sabotage. This option could be pursued using conventional 230 kV transmission lines but would require a much greater number of lines because of the conventional lines' reduced capacity, and this option might also require additional equipment, such as series capacitors, to reduce the impedance of the conventional transmission lines to attract sufficient power flow.

We note that the HTS cables shown in Figure 3.2 are from 30 to 75 miles long. Although such cables are possible in principle, they are not feasible in the near-term for two reasons:

1. HTS cables this long would require cooling stations at periodic intervals, for example at each mile or two, similar to today's underground oil-filled pipe-type transmission cables. Although existing cryocooler technology[14] could be used in this application, the cooling stations still need to be designed, integrated with the cable technology, and then demonstrated at the appropriate voltage, ampacity, and length.

2. The amount of HTS conductor required for such long cable lengths exceeds the output of current manufacturing facilities. Thus, the 230 kV HTS cables included in these Path 15 simulations most likely will not be available for many years. However, the demonstration with both the Chicago and Path 15 simulations of parallel paths using HTS cables that increase reliability, or increase transfer capacity while maintaining the same level of reliability, remains both valid and generalizable. As HTS cable demonstrations continue to increase in voltage and length,[15] the benefits of parallel paths using HTS cables will be applicable to an increasing range of grid situations.

---

[14]Radebaugh, Ray, "Development of the Pulse Tube Refrigerator as an Efficient and Reliable Cryocooler," submitted to the Proceedings Institute of Refrigeration (London, England) 1999–2000, National Institute of Standards and Technology, Gaithersburg, Md., 2000.

[15]"Demonstration of a Pre-Commercial Long-Length Superconducting Cable System Operating in the Power Transmission Network on Long Island, N.Y." and "Long Length High Temperature Superconducting Power Cable to Be Operated at a Columbus, Ohio Substation," U.S. Department of Energy, Washington, D.C., press release, September 24, 2001.

4. Engineering Comparisons Between HTS Cables and Conventional Transmission Cables and Conductors

Chapter 3 demonstrated that using HTS cables can provide reliability and transfer capacity benefits in specific transmission grid situations. This chapter investigates the conditions under which such HTS cable usage may also provide energy savings and concomitant life-cycle cost benefits. In this chapter, we describe the comparison between HTS cables and conventional underground cables at 138 kV, such as those in the downtown Chicago simulation described in Chapter 3, and the comparison between HTS cables and conventional overhead conductors at 230 kV, such as those in the California Path 15 simulation.

These comparisons are based on existing data on the engineering and cost characteristics of the conventional cables and conductors and data and engineering estimates for the HTS cables derived from experience with demonstration systems.[1] The analysis shows that there is a minimum length for HTS cables to provide energy savings and life-cycle cost benefits, and for maximum benefits they should be installed in situations in which their utilization is as close to full capacity as possible. This result is obtained because HTS cables require power consumption for cooling whether or not current flows through them.

Comparison Between HTS Cables and 138 kV Conventional Underground Cables

In this section, we draw comparisons between HTS and conventional underground cables. First, we discuss our basic analysis and its limitations, and then present a comparison of energy loss and life-cycle cost estimates for HTS and conventional underground cables.

---

[1]See Appendix A for a description of the power technologies that are currently being demonstrated.

Description and Limitations of the Analysis

The cables that were identified in Chapter 3 in the downtown Chicago simulations as candidates for replacement by HTS cables operate at 138 kV or 345 kV with a maximum power rating between 60 MVA and 1,360 MVA. Most of the cables we identified operate at 138 kV with a maximum power rating between 60 MVA and 317 MVA. In the comparisons that follow, we consider a 138 kV cross-linked polyethylene (XLPE) insulated copper conductor cable with a maximum power rating of 258 MVA as a representative conventional cable. For the HTS cable, we use the same cable that was used in the simulation, a Pirelli demonstration cable that operates at 132 kV with a maximum power rating of 685 MVA. The higher power rating of the HTS cable translates in retrofit applications, such as the simulation for downtown Chicago, into increased transfer capacity without requiring new conduits. Current field tests are demonstrating the necessary on-site cable pulling and splicing technology to enable such retrofits while using existing conduits.[2]

In the life-cycle cost comparison, we include the manufacturing cost of the HTS and conventional cables and the costs associated with the refrigeration system required to keep the HTS cable at its operating temperature. We do not include the cost of deploying the underground cables because the cost will be site specific, although we recognize that HTS cable installation will most likely be used in situations in which the additional power capacity of the HTS cable will reduce excavation costs, as in the downtown Chicago example. We take into account the higher maximum power rating of the HTS cable by computing energy requirements per unit of delivered power. Implicit in this choice is the assumption that the HTS cable will be employed in an application that uses its increased power-carrying capability.

The parameters of the HTS cable used in this analysis are representative but may not correspond to real cables in all respects. This limitation exists because current HTS cables are demonstration units and the values of key parameters may be different for commercial units. Accordingly, data and engineering estimates reflect present systems and are based on available literature. Thus, the analysis provided here should be regarded as a framework that can be used to make more-detailed cable comparisons as more-precise information becomes available. We also emphasize that energy savings and life-cycle cost benefits described in this chapter are not the sole or even the most important benefits of HTS cable use.

---

[2]Kelley, Nathan, and Jon Jipping, "HTS Cable System Demonstration at Detroit Edison," IEEE Power Engineering Society Winter Meeting, January 2002, available on CD from the Institute of Electrical and Electronics Engineers (www.ieee.org).

As noted previously, the key driving forces for the development of HTS cables are the increased reliability and power transfer capacity that can result from their use.[3]

Energy Loss Comparison of HTS and XLPE Cables

AC conduction losses in high-temperature superconductors are significantly lower than those with XLPE cables.[4] However, for an HTS cable to dissipate less energy than an XLPE cable during operation, the improvement in energy dissipation of the HTS cable must exceed the energy required to maintain its operating temperature. Later in this section, we investigate the operating parameters of the HTS cable and its cooling apparatus to determine whether its energy dissipation is less than that of existing cables. For the purpose of this analysis, we consider a three-phase HTS cable with a capacity to carry 3,000 amperes (A) at 132 kV.[5] For comparison, we consider a three-phase XLPE insulated copper conductor cable with a capacity to carry 1,080 A at 138 kV.[6] The small difference in voltage reflects different standards in the United States and Europe and is not consequential to this analysis.

The flow of an AC current through a conductor results in conduction losses that are proportional to the conductor's length. One critical difference between the superconducting and conventional cases is that the heat generated by the XLPE cable is dissipated by the surrounding medium, whereas heat generated in the superconductor must be actively removed to maintain the HTS cable operating temperature.[7] In addition to conduction losses and the associated heating, the HTS cable heat load also includes thermal losses from heat leakage through the cable thermal insulation and the terminations at which the HTS cable must connect to system elements at ambient temperature.

---

[3]The simulations in Chapter 3 demonstrate that such benefits are possible. See also Chang, Kenneth, "High Temperature Superconductors Find a Variety of Uses," *New York Times,* May 29, 2001.

[4]See, for example, Marsh, G. E., and A. M. Wolsky, *AC Losses in High-Temperature Superconductors and the Importance of These Losses to the Future Use of HTS in the Power Sector*, Argonne National Laboratory, Argonne, Ill., May 18, 2000, pp. II-3–II-4.

[5]The parameters we used to characterize the HTS cable are representative of HTS cables demonstrated in Europe by Pirelli, as described by Andrea Mansoldo, Pierluigi Ladie, Marco Nassi, Christopher Wakefield, Antonio Ardito, Paula Bresesti, and Sergio Zanella in "HTS Cables Technologies and Transmission System Reliability Studies," presented at the IEEE Power Engineering Society Summer Power Meeting, Seattle, July 2000.

[6]*Underground Transmission Cable High Voltage (HV) and Extra High Voltage (EHV) Crosslinked Polyethylene (XLPE) Insulation Extruded Lead or Corrugated Seamless Aluminum (CSA) Sheath 69-345 kV, Copper or Aluminum Conductor*, BICC Cables Company, West Nyack, N.Y.

[7]Some underground cables are actively cooled by flowing oil that must be pumped through the system. The energy requirements and cost of such systems is not included in our analysis.

42

A cryogenic refrigerator, or cryocooler, may be used to remove this heat load and maintain the HTS cable at its operating temperature of 65 K (–208°C). Existing cryocoolers operating at the temperature of HTS cables require many times more input power than the heat they remove at the operating temperature. For example, the commercial Gifford-McMahon–type AL300 cryocooler manufactured by Cryomech, Inc.[8], requires nearly 26 W (watts) of input power for every watt of cooling that it provides at 65 K. A number of Stirling- and pulse-tube–type cryocoolers with 100-1,000 W heat removal capability at 80 K are somewhat more efficient. For example, the Model 20K industrial gas liquifier under development by Mesoscopic Devices[9] requires 20 W of input power for each watt of cooling.

Based upon experimental data, the conduction loss for HTS cables with current capability up to 2,500 A is estimated to be 1 watt per meter (W/m) for each phase.[10] Termination losses are estimated at 300 to 350 W per termination for each phase,[11] based upon measurements made on a 5 meter (m) single-phase prototype version of the 30 m three-phase HTS demonstration cable that is currently powering the Southwire manufacturing complex in Carrolton, Georgia.[12] The HTS cable thermal losses will depend on the design of the cable and its envelope that contains the cooling fluid. An estimate of 0.5–1.5 W/m per phase was made based upon vacuum insulation with thermal conductivity of 0.1–0.2 mW/(m·K).[13] A range of 0.6–5.0 W/m was suggested based upon

[8]Specifications are available at http://www.cryomech.com.

[9]Specifications are available at http://www.mesoscopic.com/cryogenics.htm.

[10]Willis, Jeffrey O., David E. Daney, Martin. P. Maley, Heinrich J. Boenig, Renata Mele, Giacomo Coletta, Marco Nassi, and John R. Clem, "Multiphase AC Loss Mechanisms in HTS Prototype Multistrand Conductors," *IEEE Transactions on Applied Superconductivity*, Vol. 11, No. 1, March 2001, p. 2189; Nassi, Marco, Pierluigi Ladie, Paola Caracino, Sergio Dpreafico, Giorgio Tontini, Michel Coevoet, Pierre Manuel, Michele Dhaussy, Claudio Serracane, Sergio Zannella, and Luciano Martini, "Cold Dielectric (CD) High-Temperature Superconducting Cable Systems: Design, Development, and Evaluation of the Effects on Power Systems," manuscript received by RAND September 19, 2000; Marsh, G. E., and A. M. Wolsky, *AC Losses in High-Temperature Superconductors and the Importance of These Losses to the Future Use of HTS in the Power Sector*, Argonne National Laboratory, Argonne, Ill., May 18, 2000, pp. II-3–II-4.

[11]Marsh, G. E., and A. M. Wolsky, *AC Losses in High-Temperature Superconductors and the Importance of These Losses to the Future Use of HTS in the Power Sector*, Argonne National Laboratory, Argonne, Ill., May 18, 2000, pp. II-3–II-4.

[12]Gouge, M. J., J. A. Demko, P. W. Fisher, C. A. Foster, J. W. Lue, J. P. Stovall, U. Sinha, J. Armstrong, R. L. Hughey, D. Lindsay, and J. Tolbert, "Development and Testing of HTS Cables and Terminations at ORNL," *IEEE Transactions on Applied Superconductivity*, Vol. 11, No. 1, March 2001, p. 2352.

[13]Oestergaard, Jacob, Jan Okholm, Karin Lomholt, and Ole Toennesen, "Energy Losses of Superconducting Power Transmission Cables in the Grid," *IEEE Transactions on Applied Superconductivity*, Vol. 11, No. 1, March 2001, p. 2376.

experience with industrial cryogenic cooling systems;[14] this is roughly consistent with the range projected by Southwire and Pirelli for planned HTS cables.[15]

For the XLPE cable, the conduction losses are by far the dominant losses, and the only losses included in our analysis, and can be computed from the geometry of the cable and the resistivity of the conductor. For the XLPE cable, the conduction loss equals 26.3 W/m.[16] In Table 4.1, we list the HTS and XLPE cable parameters.

It is clear from Table 4.1 that the termination loss overwhelms conduction and thermal losses for short HTS cables. In fact, HTS cable losses exceed those in the XLPE cable for lengths less than approximately 30 m. In Appendix B, we develop expressions for the energy losses in both cable types for the quantities shown in Table 4.1, the coefficient of performance of the cryocooler (ratio of input power to heat removed) and the level of utilization of the cables (i.e., the fraction of their maximum rating at which they are operated). Equating these expressions provides a parametric relationship for the HTS cable to achieve energy savings. This relationship determines the minimum HTS cable length for HTS cable energy savings.

Using the parameters in Table 4.1 and the characteristics of the AL300 cryocooler described earlier, we find that the minimum length of the HTS cable for energy savings is 937 m. The simulations of the downtown Chicago transmission grid described in Chapter 3 identified 138 kV conventional cable candidates for

Table 4.1

Operating Characteristics of HTS and XLPE Cables

	HTS	XLPE
Voltage (kV)	132	138
Maximum current (A)	3,000	1,080
Thermal loss per phase (W/m)	1–5	N/A
Termination loss per phase (W)	650	N/A
Conduction loss per phase (W/m)	1	26.3
Maximum power rating (MVA)	685	258
Operating temperature (K)	65	Ambient

---

[14]Gerhold, J., "Power Transmission," in B. Seeber, ed., *Handbook of Applied Superconductivity*, Vol. 2, Bistol, UK: Institute of Physics Publishing, 1998, p. 1685.

[15]Marsh, G. E. and A. M. Wolsky, *AC Losses in High-Temperature Superconductors and the Importance of These Losses to the Future Use of HTS in the Power Sector*, Argonne National Laboratory, Argonne, Ill., May 18, 2000, pp. II-3–II-4.

[16]This number is based on the operation of a copper conductor cable manufactured by BICC carrying 138 kV and 1,080 A.

44

retrofit by HTS cable as being between 0.6 miles and 2.2 miles in length, which is greater than the 937 m minimum length.

Appendix B shows that the following expression determines the requirements on cryocooler performance and cable utilization for the HTS cable to achieve energy savings:

$$p(\theta + \omega_s) \neq \omega R - \omega_c u_s, \tag{4.1}$$

where ρ is the coefficient of performance of the cryocooler, θ is the thermal loss per phase of the HTS cable, ω is the conduction loss per phase of the HTS or XLPE cable, u is the utilization of the cable (discussed later in this section), R is the ratio of MVA power ratings, and the subscripts denote superconductor (s) or conventional (c).

We define *utilization* as the fraction of the maximum rated power carried by the cable. Because the conduction loss is proportional to the current squared for the XLPE cable (Ohm's Law) and to the current cubed for the HTS cable,[17] different time variations of utilization can lead to different values of loss. It is customary to define a load curve that represents the variation in utilization over a given time period, such as a day, and then to integrate the appropriate function of the current using this curve to determine the loss. For the purposes of this analysis, we adopted the simplified procedure of assuming that the current flows at its maximum level for a fraction of the time and does not flow at all the rest of the time, hence the linear dependence on u in Equation (4.1).[18]

The performance of any refrigerator, including the HTS cryocooler, is customarily measured by comparing that performance to the best performance that can be achieved when the refrigerator is operating between its maximum and minimum temperatures. This maximum possible efficiency is called the "Carnot efficiency," which, according to the second law of thermodynamics, is determined solely by these two temperatures.[19]

Figure 4.1 shows the relationship between the HTS cryocooler performance, as a percentage of Carnot efficiency, and the HTS utilization, as obtained from Equation (4.1). The cryocooler efficiency is the inverse of the coefficient of

---

[17]Gerhold, J., "Power Transmission," in B. Seeber, ed., *Handbook of Applied Superconductivity*, Vol. 2, Bistol, UK: Institute of Physics Publishing, 1998, p. 1636.

[18]Alternatively, one could use the instantaneous value of the current, in which case u^2 would appear in the ω_c term and u^3 in the ω_s terms in Equation (4.1). The difference between the two approaches is negligible for high utilization (e.g., $u > 0.6$) and increases in importance as the utilization decreases.

[19]See, for example, Fermi, Enrico, *Thermodynamics*, New York: Dover Publications, Inc., 1936, Chapter III.

performance, and all of the other parameters except utilization are as shown in Table 4.1. The three curved lines in Figure 4.1 represent different values of the HTS cable thermal loss per phase within the range shown in Table 4.1.

The curved lines in Figure 4.1 also satisfy Equation (4.1) and thus define the boundaries between values of cryocooler efficiency and cable utilization for which the HTS cable uses less energy than the XLPE cable (i.e., the region above the curve) and for which the HTS cable uses more energy than the XLPE cable (i.e., the region below the curve). It is clear from Figure 4.1 that for any utilization there is a minimum value of cryocooler efficiency for the HTS cable to achieve energy savings, and that this value increases as the utilization decreases. This is consistent with the observation that the cryocooler uses power to maintain the HTS cable operating temperature regardless of whether or not the cable is carrying current. Similarly, the minimum cryocooler efficiency for energy savings increases as the thermal loss increases because the cryocooler heat load increases.

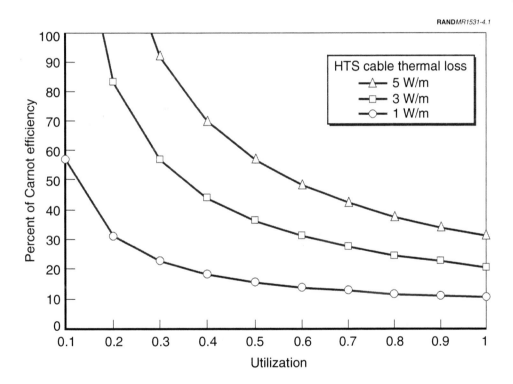

Figure 4.1—Boundaries Showing that HTS Cables with Cryocooler Efficiency and Utilization Values Above the Curves Use Less Energy Per Unit of Power Delivered than the XLPE Cable

The AL300 cryocooler described earlier requires 7,400 W of input power to remove 285 W of heat load at 65 K,[20] which translates to 14 percent of Carnot efficiency. A number of Stirling-type and pulse-tube–type cryocoolers with 100–1,000 W heat-removal capability at 80 K have achieved efficiencies between 15 and 20 percent of Carnot.[21] Thus Figure 4.1 indicates that in order to achieve energy savings with existing cryocooler technology it is important to engineer the HTS cable so that its thermal loss per phase is within the lower portion of the range shown in Table 4.1, or near 1 W/m. It is also important to operate the HTS cable at high utilization, for example $u > 0.6$. Within this parameter space, HTS cables will use less energy than conventional XLPE cables per unit of delivered power.

Life-Cycle Cost Estimates for HTS and XLPE Cables

The previous section demonstrated that there is a range of parameter values and operating conditions under which an HTS cable dissipates less energy per unit of delivered power than an XLPE cable does. We now consider the life-cycle cost of owning and operating the two systems. The HTS cable must not only be more energy efficient than the cable that it replaces, it must also save enough energy over its expected lifetime to compensate for the higher acquisition cost of the HTS cable and its cooling equipment. Our model includes losses incurred when current flows in the cables and the cost incurred to keep the HTS cable at its operating temperature. The operating parameters of the cables are as stated earlier with an expected lifetime of 40 years.[22]

It is difficult to estimate the manufacturing cost of commercial HTS cables because only prototype and demonstration cables have been produced. Because the cost of the demonstration cables is dominated by the cost of the HTS wire (currently $200 per kAm),[23] we use the HTS wire cost as a proxy for the cost of the manufactured cable. It is anticipated that HTS cable cost will be lower than this proxy value as a result of projected reductions in HTS wire cost from the DOE-sponsored government-industry Superconductivity Partnership Initiative

---

[20]Specifications are available at http://www.cryomech.com.

[21]Radebaugh, Ray, "Development of the Pulse Tube Refrigerator as an Efficient and Reliable Cryocooler," submitted to the Proceedings Institute of Refrigeration (London, England) 1999–2000, National Institute of Standards and Technology, Gaithersburg, Md., 2000.

[22]This value was based on the expected lifetime of electrical cables suggested by Mulholland, Joseph, Thomas P. Sheahen, and Ben McConnell, *Analysis of Future Prices and Markets for High Temperature Superconductors* (draft), U.S. Department of Energy, September, 2001, http://www.ornl.gov/HTSC/pdf/Mulholland%20Report.pdf (last accessed April 1, 2002).

[23]Lyons, Chet, private communications, American Superconductor.

(SPI).[24] However, the manufactured HTS cable will use more wire than the ratio of kAm to the cable rating in kA because the wire is wound in a spiral. The estimated cost of the XLPE cable is $100 per kAm.[25]

The operation of a superconducting cable requires refrigeration equipment. We estimate the cost of a refrigerator by calculating the power necessary to maintain the cable at its operating temperature and multiplying this by the cost of a cryogenic refrigerator of a given power output.[26]

The operating costs are calculated by multiplying the calculated energy dissipated for each cable or conductor by the price of electricity. The appropriate price of electricity depends upon the operating conditions of the cable or conductor. If the cable or conductor is used in a situation in which the transmitted power is fixed, then any loss savings will reduce production requirements, and the appropriate price is the production cost, or wholesale price. However, if the cable or conductor is able to transmit additional power because of reduced losses, then the appropriate price is the marginal price of electricity. As a proxy, we use the average retail price of electricity.

By totaling the costs associated with the purchase and operation of each cable or conductor, we are able to estimate the lifetime cost of the HTS cable and the XLPE cable. The operating costs are discounted over the operating period at a rate of 7 percent per year. We note that the operating cost of HTS cables may be somewhat higher than the estimate, because we have not included maintenance and replacement costs for the refrigeration system. On the other hand, the capital cost estimate for the refrigeration system is based upon current low-volume manufacturing, so that it is likely too high.

We estimate the cost of operating the HTS cable and the XLPE cable for electricity costs of 5¢ per kilowatt hour (kWh) and 10¢ per kWh. This sets a reasonable range because average electric utility prices for 2000 were 4.45¢ per kWh for industrial customers, 7.2¢ per kWh for commercial customers and 8.21¢ per kWh

[24]American Superconductor Corporation, the manufacturer of the BSCCO wire used in the demonstration cables has projected a price reduction to $50 per kAm based upon the anticipated opening of a new manufacturing facility in 2004. The DOE has set a target of $10 per kAm for a second-generation wire based upon a YBCO coated conductor technology under development. A recent survey of 70 experts carried out for DOE put the projected cost of second-generation HTS wire in the range of $10 to $50 per kAm. See *Proceedings Coated Conductor Development Roadmapping Workshop—Charting Our Course*, U.S. Department of Energy Superconductivity for Electric Systems Program, January, 2001.

[25]The cost of the XLPE cable is an estimate by Donald Von Dollen, EPRI (formerly the Electric Power Research Institute), private communication, which is consistent with cable auction costs found at http://www.piap.ch/auction/wire.html.

[26]For this analysis, we set the cost of cryogenic equipment at $100,000 per kW of heat load, which is consistent with the cost of the commercial AL300 cryocooler described earlier.

48

for residential customers.[27] As we did in the previous section, we compare costs of the HTS cable and XLPE cable per unit of delivered power.

Appendix B describes a procedure that results in an expression relating the HTS and XLPE cable parameters, analogous to Equation (4.1), for the HTS cable to have a life-cycle cost benefit per unit of delivered power over its 40-year lifetime with the 7 percent discount rate. Figure 4.2 shows the relationship between HTS cable cost and cryocooler efficiency to achieve a HTS cable life-cycle cost benefit for the range of thermal loss values previously discussed (θ = 1-5 W/m),[28] with an electricity cost of 10¢ per kWh. Figure 4.3 shows the same relationship with an electricity cost of 5¢ per kWh. For all values of HTS cable parameters that fall *below* any curve in both figures, the HTS cable will have a life-cycle cost benefit (i.e., the HTS cable will have a lower life-cycle cost per unit of power delivered than the XLPE cable).

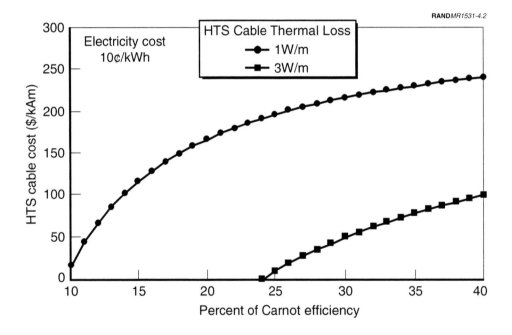

Figure 4.2—Boundaries Showing that HTS Cables with Cable Cost and Cryocooler Efficiency Values Below the Curves Have a Lower Life-Cycle Cost Per Unit of Power Delivered than the XLPE Cable, Electricity Cost of 10¢ per kWh

---

[27]Energy Information Administration, *Monthly Electric Utility Sales and Revenue Report with State Distributions*, "U.S. Electric Utility Average Revenue per Kilowatthour by Sector, 1990 Through December 2000," U.S. Department of Energy, available at http://www.eia.doe.gov/cneaf/electricity/epm/epmt52.txt (last accessed March 28, 2002).

[28]Thermal loss of 5 W/m is not shown in the figure because there was no life-cycle cost benefit for any value of HTS cable cost or cryocooler efficiency for this case.

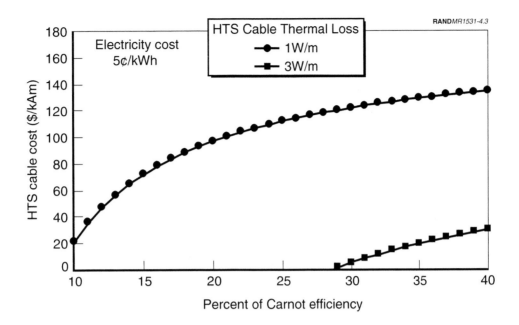

Figure 4.3—Boundaries Showing that HTS Cables with Cable Cost and Cryocooler Efficiency Values Below the Curves Have a Lower Life-Cycle Cost Per Unit of Power Delivered than the XLPE Cable, Electricity Cost of 5¢ per kWh

Figures 4.2 and 4.3 illustrate how minimizing thermal losses carries over to the life-cycle cost analysis. There are no values of the parameters for which the HTS cable has a lower life-cycle cost than the XLPE cable when the thermal loss equals 5 W/m, and only for very low HTS cable cost and very high cryocooler efficiency is there a life-cycle cost benefit when compared with the XLPE cable when the thermal loss equals 3 W/m.

For the minimum thermal loss case of 1 W/m, HTS cables show a life-cycle cost benefit compared with an XLPE cable per unit of delivered power when the cost of the HTS cable is below $200 per kAm and the efficiency of the cryocooler is greater than 25 percent Carnot[29] with an electricity cost of 10¢ per kWh. These requirements are not inconsistent with estimates based upon the anticipated reduction of HTS wire cost described earlier and recent trends in the development of large cryocoolers.[30]

For the minimum thermal loss case of 1 W/m and an electricity cost of 5¢ per kWh, the requirement for an HTS cable life-cycle cost benefit is a manufactured

---

[29]A 25 percent Carnot efficiency means that, for a cryocooler operating at 65 K, 14.3 watts of electric power must be supplied for each watt of cooling obtained.

[30]Radebaugh, Ray, "Development of the Pulse Tube Refrigerator as an Efficient and Reliable Cryocooler," submitted to the Proceedings Institute of Refrigeration (London, England) 1999–2000, National Institute of Standards and Technology, Gaithersburg, Md., 2000.

50

cable cost near $100 per kAm for a cryocooler efficiency greater than 25 percent Carnot. In this case, the HTS cable manufacturing cost must be nearly equal to that of the XLPE cable, which will be difficult to achieve unless the cost of HTS wire goes below that of copper wire.

We conclude from Figures 4.2 and 4.3 that the life-cycle cost comparison between HTS and XLPE cables is sensitive to the cost of electricity, with the HTS cable being cheaper for 10¢ per kWh and the XLPE cable being cheaper for 5¢ per kWh. The fact that these life-cycle cost estimates that are based upon acquisition and operating costs are so close to each other suggests that the site-specific costs of deployment which we did not include in the calculation can be important. For example, Figures 4.2 and 4.3 demonstrate that the life-cycle costs of HTS and XLPE cables are close enough that in situations in which the HTS cable's higher power capacity can eliminate excavation costs (e.g., in downtown Chicago) the HTS cable may also be the lowest-cost option.

The minimum length at which HTS cables provide life-cycle cost benefits can be computed for specific values of cyrocooler efficiency and thermal loss. As in the previous paragraph, we consider the minimum thermal loss case. For example, with an electricity cost of 10¢ per kWh, an HTS cable cost of $175 per kAm, and cryocooler efficiency equal to 27 percent of Carnot, the minimum HTS cable length for the life-cycle cost benefit is 1,660 m. With an electricity cost of 5¢ per kWh, an HTS cable cost of $90 per kAm, and cryocooler efficiency equal to 27 percent of Carnot, the minimum HTS cable length for the life-cycle cost benefit is 1,260 m. These cable lengths are similar to those of the HTS cable retrofits in the downtown Chicago power-flow simulations.

Two parameters used in the previous analysis may be varied to reflect operational conditions and commercial developments. The analysis assumed full utilization, which may not properly reflect operating conditions. As utilization decreases from unity, the minimum length and minimum cryocooler efficiency necessary to achieve a life-cycle cost benefit increase. The required cable cost for the HTS cable to provide a life-cycle cost benefit also decreases with decreased utilization.

Another factor that is expected to influence the minimum cost of HTS cable for the life-cycle cost benefit is the cost of the cryocooler, which increases the acquisition cost of the HTS system. If HTS cables become a commercial product, it is likely that cryocooler cost will be reduced according to a typical manufacturers' learning curve as production volume increases.[31] Reduced cryocooler cost would increase the HTS cable cost and decrease the cryocooler efficiency at which the HTS cable will show a life-cycle cost benefit. Thus, the two parameters (utilization and cryocooler cost) introduce effects that work in opposite directions. This is illustrated in Figure 4.4, which covers utilization ranging from 0.5 to 1.0 and cryocooler costs of $10 per W to $100 per W for a thermal loss of 1 W/m, with an electricity cost of 5¢ per kWh. We note that the two effects approximately cancel out for $u = 0.5$ and a cryocooler cost of $10 per W. Because utilization will depend upon the operating conditions of the specific application, and cryocooler cost will depend upon the ultimate production volume, we will continue to use $u = 1$ and the current cryocooler cost of $100 per W in the following analysis.

The minimum lengths described earlier for the life-cycle cost benefit of HTS cables were calculated under the assumption of a 40-year operating lifetime. This operating lifetime corresponds to a 40-year payback time for the operational cost savings of the HTS cable as compared with the difference in acquisition cost between the HTS cable plus the cryocooler and the XLPE cable. Because the conduction loss of the XLPE cable increases with cable length, longer HTS cables will provide payback in a shorter period of time. Appendix B derives an expression for the discounted payback time as a function of HTS cable length. Figure 4.5 shows this relationship for two scenarios. In one case, the electricity cost is 10¢ per kWh and the HTS cable cost is $175 per kAm. In the other case, the electricity cost is 5¢ per kWh and the HTS cable cost is $90 per kAm (which will be difficult to achieve, as noted earlier). The cryocooler is assumed to operate at 27 percent of Carnot efficiency. In both cases, for a cable length much greater than the minimum length, the discounted payback rapidly approaches an asymptotic value, indicated by the dotted lines in Figure 4.5.

Figures 4.6 and 4.7 show the discounted payback time for long cables (i.e., the asymptotic value) as a function of HTS cable cost and cryocooler efficiency. For an electricity cost of 10¢ per kWh, discounted payback periods of between 10 and 40 years can be obtained for HTS cable costs and cryocooler efficiencies in the range of engineering projections, whereas for an electricity cost of 5¢ per kWh, difficult-to-achieve HTS cable cost and cryocooler efficiency targets are required for a discounted payback of less than 40 years.

---

[31]Nisenoff, M., "Status of DoD and Commercial Cryogenic Refrigerators," *Cryogenics Vision Workshop for High-Temperature Superconducting Electric Power Systems Proceedings*, U.S. Department of Energy Superconductivity Program for Electric Systems, July 1999.

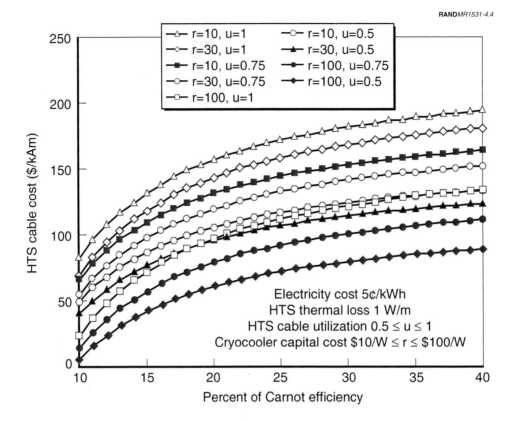

RAND*MR1531-4.4*

Figure 4.4—Boundaries Showing that HTS Cables with Cable Cost and Cryocooler Efficiency Values Below the Curves Have a Lower Life-Cycle Cost Per Unit of Power Delivered than the XLPE Cable, Various Utilization and Cryocooler Cost Values, Electricity Cost of 5¢ per kWh, HTS Cable Thermal Loss of 1 W/m

Comparison Between HTS Cables and 230 KV Conventional Overhead Lines

In this section, we draw comparisons between HTS cables and conventional overhead transmission lines. First, we discuss our basic analysis and its limitations, and then present a comparison of energy loss and life-cycle cost estimates for HTS cables and conventional overhead lines.

Description and Limitations of the Analysis

The power lines identified in the California Path 15 simulations (Figure 3.2 in Chapter 3) as candidates for replacement by HTS cables operate at 230 kV with a maximum power rating of about 300 MVA. In the comparison that follows, we consider a 230 kV aluminum conductor steel reinforced (ACSR) overhead line

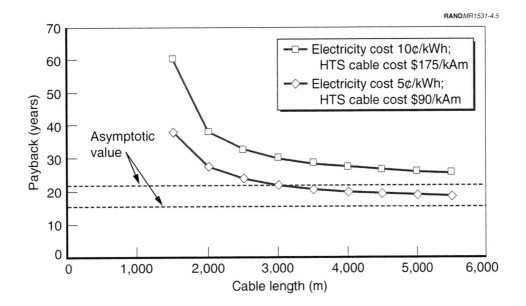

Figure 4.5—Discounted Payback Time Versus Cable Length for HTS Cable Compared
with XLPE Cable, Cryocooler Efficiency 27 Percent of Carnot

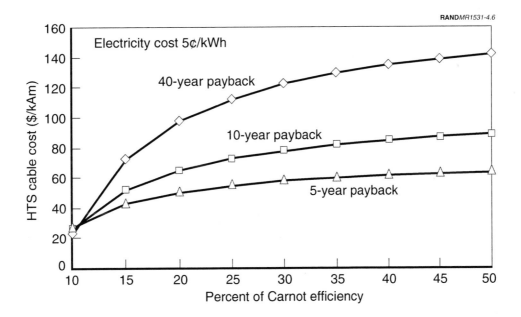

Figure 4.6—Discounted Payback Time for HTS Cable Compared with XLPE Cable as a
Function of HTS Cable Cost and Cryocooler Efficiency, Electricity Cost of 5¢ per kWh

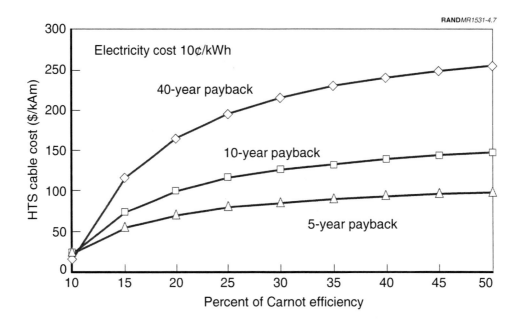

Figure 4.7—Discounted Payback Time for HTS Cable Compared with XLPE Cable as a Function of HTS Cable Cost and Cryocooler Efficiency, Electricity Cost of 10¢ per kWh

with a maximum power rating of 311 MVA as a representative conventional overhead line. For the HTS cable, we use the same cable that was used in the simulation, a proposed Pirelli cable that operates at 230 kV with a maximum power rating of 1,195 MVA. The higher power rating of the HTS cable can be used, as in the Path 15 simulations, to increase transfer capacity.

In the life-cycle cost comparison that follows, we follow the same pattern as outlined in the comparison of the HTS and XLPE cables earlier in this chapter. We include the manufacturing cost of the HTS cable and the conventional line and the costs associated with the refrigeration system for the HTS cable. We do not include the cost of deploying the HTS underground cable and the ACSR overhead line, as discussed later. As in the HTS-XLPE comparison, we take into account the higher maximum power rating of the HTS cable by computing energy requirements per unit of delivered power, implicitly assuming that the HTS cable will be employed in an application that uses its increased power-carrying capability.

As noted in Chapter 2, the overwhelming majority of the U.S. transmission grid consists of overhead lines, which are much less expensive to construct than underground cables. Thus, we expect that the overhead ACSR transmission line will be deemed preferable to the HTS (or any) underground cable unless there is a site-specific cost problem or other problem such as high land-use demands or right-of-way costs, difficulty in obtaining necessary siting approvals, or interest in increasing power transfer at a lower voltage, such as was simulated for Path 15. The comparisons of energy losses and life-cycle costs for HTS cables and ACSR lines, presented next, apply to the case in which such site conditions mitigate the ACSR deployment cost advantage, which would otherwise dominate any HTS operating cost savings. As with the HTS-XLPE comparison, we also recognize that decisions to use HTS cables will likely be based upon reliability and transfer capacity benefits rather than energy and life-cycle cost savings.

Energy Loss Comparison of HTS Cables and Overhead ACSR Transmission Lines

In this analysis, we consider a three-phase ACSR transmission line with a capacity to carry 780 A at 230 kV. The maximum power rating of this line is 311 MVA. This compares with 3,000 A at 230 kV, with a maximum power rating of 1,195 MVA, for the HTS cable. As for the comparison between the HTS and XLPE cables, the conduction loss for the HTS cable is estimated to be 1 W/m for each phase and termination losses are estimated at 300 to 350 W per phase. For the ACSR line, the conduction losses are by far the dominant losses, and the only losses included in our analysis. For the ACSR line, the conduction loss equals 64 W/m.[32] In Table 4.2, we list the HTS cable and ACSR line parameters.

The higher power dissipation of the ACSR line, as compared with the XLPE cable, reduces the minimum length for HTS cable energy savings, taking into account the energy use of the cryocooler to 88 m. Thus, from the point of view of energy use of HTS cables compared with an overhead line, even very short lengths of HTS cables are an attractive option. Whereas the Path 15 simulations identified retrofit candidates that were 230 kV lines approximately 40 miles long, much longer than any current or planned HTS cable demonstrations, parallel HTS cable paths to increase power transfer capacity may be possible at shorter lengths in other transmission congestion locations.

The principal benefit of HTS cables is their higher power-carrying capacity (a factor of 3.84 by comparison with the ACSR line), without increasing voltage, adding circuits, or doing both. This benefit will be most important where there are space constraints or a lack of necessary equipment for uprating voltage.

---

[32]This number is based on the operating parameters of a Southwire ACSR 636 conductor. Specifications are available at http://www.southwire.com.

Table 4.2

Operating Characteristics of HTS Cable and ACSR Line

	HTS	ACSR
Voltage (kV)	230	230
Current (A)	3,000	780
Thermal loss per phase (W/m)	1–5	N/A
Termination loss per phase (W)	650	N/A
Conduction loss per phase (W/m)	1	64
Maximum power rating (MVA)	1,195	311
Operating temperature (K)	65	Ambient

The comparison of energy use between the HTS cable and the ACSR line depends upon utilization and cryocooler efficiency according to Equation (4.1), which is also derived in Appendix B. The curved lines in Figure 4.8 (which is analogous to Figure 4.1) satisfy Equation (4.1) for the parameters of the ACSR line and the HTS cable shown in Table 4.2, and thus define the boundary between the values of cryocooler efficiency and cable utilization for which the HTS cable uses less energy than the ACSR line (the region above each curved line in the figure) and for which the HTS cable uses more energy than the ACSR line (the region below the curved lines) per unit of delivered power.

For high utilization (e.g., $u > 0.6$) and cryocooler efficiency in the 15–20 percent range, the HTS cable uses less energy per unit of delivered power than the ACSR line for the full range of thermal heat loss shown in Table 4.2. For the minimum thermal loss case (1 W/m) and the 14 percent efficiency of the AL300 cryocooler, the HTS cable shows energy savings as compared with the ACSR line for even much-lower utilization. We note that this is a consequence of the relatively high loss of the ACSR line with parameters similar to that simulated for Path 15. One could reduce this loss by using larger conductors and multiple ACSR circuits to carry the same power, assuming that the additional space was available at an acceptable cost, including the cost of obtaining siting approval.

Life-Cycle Cost Estimates for HTS Cables and ACSR Transmission Lines

The previous section demonstrated that there is a range of parameter values and operating conditions under which an HTS cable dissipates less energy per unit of delivered power than does an ACSR line. We now consider the life-cycle cost of owning and operating the two systems. As noted earlier, we disregard site-specific costs, such as purchase of a right-of-way for the ACSR line and excavation for the HTS cable, so that these comparisons will be useful for cases in which site constraints mitigate the usual cost advantage of the overhead line.

RAND*MR1531-4.8*

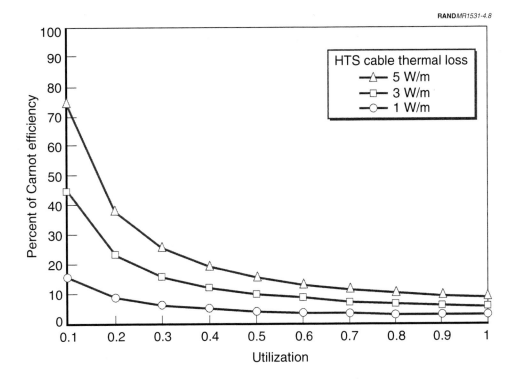

Figure 4.8—Boundaries Showing that HTS Cables with Cryocooler Efficiency and Utilization Values Above the Curves Use Less Energy Per Unit of Power Delivered than the ACSR Line

The purpose of comparing the acquisition and operating costs of ACSR lines and HTS cables is that the HTS cable must be not only more energy-efficient than the technology that it replaces, it must also save enough energy over its expected lifetime to compensate for the higher acquisition cost of the HTS cable and its cooling equipment. Our model includes losses incurred when current flows in the cable and the cost incurred to keep the cable at its operating temperature. As in the earlier comparison of the HTS and XLPE cables, we calculate the operating cost over an expected lifetime of 40 years,[33] with a discount rate of 7 percent. The operating parameters of the HTS cable and ACSR line are the same as those shown earlier. As for the XLPE comparison, we use the HTS wire cost (currently

---

[33]This value was based on the expected lifetime of electrical cables suggested by Mulholland, Joseph, Thomas P. Sheahen, and Ben McConnell, *Analysis of Future Prices and Markets for High Temperature Superconductors* (draft), U.S. Department of Energy, September, 2001, http://www.ornl.gov/HTSC/pdf/Mulholland%20Report.pdf (last accessed April 1, 2002).

$200 per kAm)[34] as a proxy for the cost of the manufactured HTS cable. The estimated cost of the ACSR cable is $6.85 per kAm.[35]

And as with the XLPE comparison, we total the costs associated with purchase and operation of the ACSR line and HTS cable to estimate the lifetime cost of the HTS cable and the ACSR line, and we estimate the cost of operating the cable and the line for electricity costs of 5¢ per kWh and 10¢ per kWh. We determine the requirements on HTS cable parameters so that the HTS cable will have a life-cycle cost benefit using the expression analogous to Equation (4.1), which is also derived in Appendix B. Figure 4.9 shows HTS cable cost and cryocooler efficiency for the range of thermal loss values previously discussed (1–5 W/m) with an electricity cost of 10¢ per kWh. Figure 4.10 shows the same solution for an electricity cost of 5¢ per kWh. For all values of HTS cable parameters that fall below the curved lines in Figures 4.9 and 4.10, the HTS cable will have a life-cycle cost benefit compared with the ACSR line per unit of delivered power.

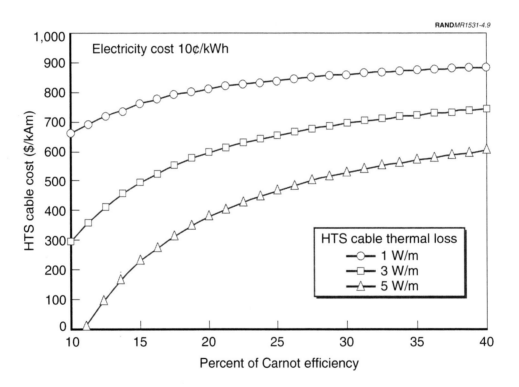

Figure 4.9—Boundaries Showing that HTS Cables with Cable Cost and Cryocooler Efficiency Values Below the Curves Have a Lower Life-Cycle Cost per Unit of Power Delivered than the ACSR Line, Electricity Cost of 10¢ per kWh

---

[34]Lyons, Chet, private communications, American Superconductor.

[35]Bare Aluminum Conductor ACSR price list, effective February 12, 2001, Alcan, Inc., Montreal, Quebec.

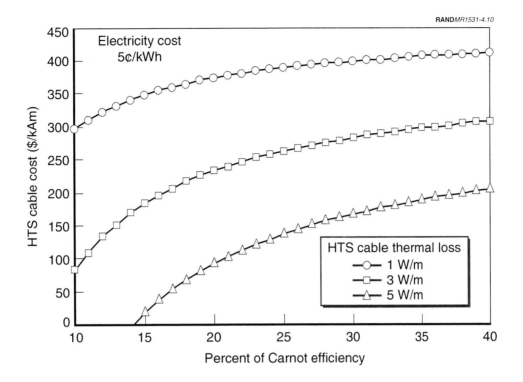

RAND*MR1531-4.10*

Figure 4.10—Boundaries Showing that HTS Cables with Cable Cost and Cryocooler
Efficiency Values Below the Curves Have a Lower Life-Cycle Cost Per Unit of Power
Delivered than the ACSR Line, Electricity Cost of 5¢ per kWh

Figures 4.9 and 4.10 illustrate the contribution of thermal loss of the HTS cable to
the life-cycle cost analysis. The higher the thermal loss, the lower the required
cost of the HTS cable for it to compete with an ACSR line on a life-cycle cost
basis. The maximum HTS cable costs are higher than the costs when the
comparison was carried out for the XLPE cable because of the higher power
dissipation of the ACSR line. For example, the HTS cable shows a life-cycle cost
benefit for the cable cost that is equal to today's HTS wire cost of $200 per kAm
and cryocooler efficiency of 20 percent Carnot for all cases illustrated in Figures
4.9 and 4.10 except for the case of electricity cost at 5¢ per kWh and HTS cable
thermal loss of 5 W/m.

The minimum length at which HTS cables provide a life-cycle cost benefit can be
computed for specific values of the parameters. For the minimum thermal loss
case (1 W/m) with an HTS cable cost of $200 per kAm and cryocooler efficiency
equal to 14 percent of Carnot (today's values), the minimum HTS cable length for
a life-cycle cost benefit is 166 m with an electricity cost of 10¢ per kWh and 391 m
with an electricity cost of 5¢ per kWh. These figures demonstrate that even short
HTS cables can provide life-cycle cost benefits per unit of power delivered as
compared with ACSR lines.

60

The analysis up to now assumed full utilization, which may not properly reflect operating conditions. As utilization decreases from unity, the minimum length and minimum cryocooler efficiency for the HTS cable life-cycle cost benefit increase and the required HTS cable cost decreases. On the other hand, the cost of cryocoolers is expected to decrease as demand increases, which would increase minimum length and minimum cryocooler efficiency and result in a higher HTS cable cost for the HTS cable life-cycle cost benefit. This pattern is illustrated in Figure 4.11, which covers utilization ranging from 0.5 to 1.0 and cryocooler costs from $10 per W to $100 per W for a thermal loss of 1 W/m and an electricity cost of 5¢ per kWh.

In comparing Figure 4.11 with Figure 4.4, we note that the two effects shown in Figure 4.11 no longer approximately cancel out. The reduced utilization has a

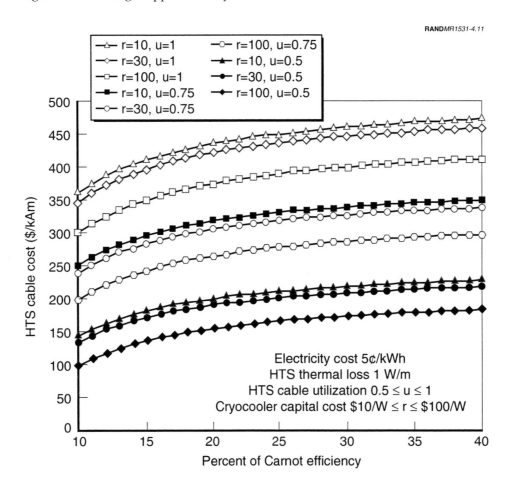

Figure 4.11—Boundaries Showing that HTS Cables with Cable Cost and Cryocooler Efficiency Values Below the Curves Have a Lower Life-Cycle Cost Per Unit of Power Delivered than the ACSR Line, Various Utilization and Cryocooler Cost Values, Electricity Cost of 5¢ per kWh, HTS Cable Thermal Loss of 1 W/m

much more significant impact for the ACSR comparison because it compensates for the higher loss of the ACSR line. However, for high utilization (e.g., $u > 0.6$), the HTS cable continues to show a life-cycle cost benefit when we assume today's values of HTS wire cost and cryocooler efficiency.

Because the conduction loss of the ACSR line increases with length, longer HTS cables will provide payback in a shorter period of time. Appendix B derives an expression for the discounted payback time as a function of HTS cable length. Figure 4.12 (which is analogous to Figure 4.5) shows the discounted payback time as a function of HTS cable length calculated from this relationship for today's values of HTS wire cost and cryocooler efficiency. For cables of lengths much longer than the minimum, the discounted payback approaches asymptotic values, as indicated by the dotted lines in Figure 4.12, of slightly less than 15 years with an electricity cost of 5¢ per kWh and slightly more than 5 years with an electricity cost of 10¢ per kWh.

Figures 4.13 and 4.14 show the discounted payback times for long cables as a function of HTS cable cost and cryocooler efficiency. Discounted payback times are slightly greater than ten years or five years for an electricity cost of 5¢ per kWh or 10¢ per kWh, respectively, for cryocooler efficiency of 14 percent Carnot and HTS cable cost of $200 per kAm.

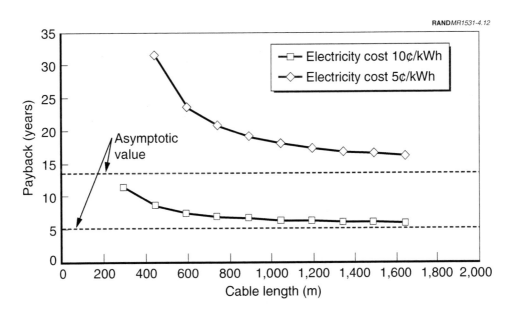

Figure 4.12—Discounted Payback Time Versus Cable Length for HTS Cable Compared with ACSR Line, HTS Cable Cost of $200 per kAm, Cryocooler Efficiency 14 Percent of Carnot

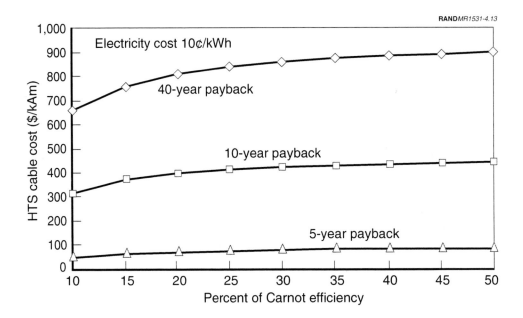

Figure 4.13—Discounted Payback Time for HTS Cable Compared with ACSR Line as a Function of HTS Cable Cost and Cryocooler Efficiency, Electricity Cost of 10¢ per kWh

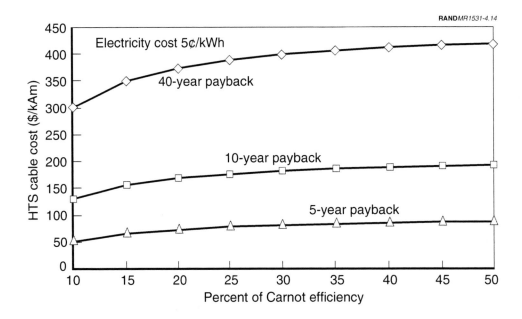

Figure 4.14—Discounted Payback Time for HTS Cable Compared with ACSR Line as a Function of HTS Cable Cost and Cryocooler Efficiency, Electricity Cost of 5¢ per kWh

5. Engineering Comparisons Between HTS and Conventional Energy Storage Systems and Transformers

This chapter describes comparisons of the energy use and life-cycle cost of HTS and conventional power technologies other than power cables. These comparisons are based on existing data on the engineering and cost characteristics of the conventional technologies and data and engineering estimates for the superconducting technologies derived from experience with demonstration and commercial systems. Separate sections discuss flywheel energy storage systems with HTS bearings, HTS transformers, and superconducting magnetic energy storage systems that use a low-temperature superconducting coil and HTS current leads.

Flywheel Energy Storage Systems

In this section, we draw comparisons between flywheel energy storage systems, including those that utilize HTS bearings, and battery energy storage systems. First, we discuss our basic analysis and its limitations and then present a comparison of energy loss and life-cycle cost estimates for these two different types of energy storage systems.

Description and Limitations of the Analysis

The increased use of consumer and commercial electronic products, most notably computers and associated hardware, calls for the delivery of high-quality power to the end user. In its study of power quality, EPRI found that most disturbances to the electrical supply last less than one second.[1] Such disturbances are particularly disruptive to the operation of electronic devices. The effects of excursions away from the rated voltage have been compiled in the Computer and Business Equipment Manufacturers Association (CBEMA) Curve.[2] (Effective

[1]EPRI Distribution Power Quality Study, cited in *Power Quality*, Trinity Flywheel Power, Livermore, Calif., http://www.trinityflywheel.com/site/apps/app_pq.html (last accessed September 20, 2000).

[2]Dorrough Electronics Inc., *CBEMA Curve*, Woodland Hills, Calif., 1999. See http://www.dorrough.com/What_s_New/PLM-120/PowerMonTM/CBEMA_Curve/cbema_curve.html (last accessed March 28, 2002).

December 1994, CBEMA changed its name to the Information Technology Industries Council [ITI].) The CBEMA Curve shows that those effects on equipment depend on the duration and extent of deviation from the desired voltage. As a result, many consumers have opted for the use of uninterruptible power supplies (UPSs).

Several approaches to UPSs exist; the operating circumstances dictate the power and energy requirements. Most UPSs today rely on batteries for energy storage. Recently, manufacturers have developed commercial FESSs to compete with these battery energy storage systems (BESSs). In the next section, we compare the energy use and life cycle cost of a BESS and a FESS. In these comparisons, we include the cost of the BESS and the FESS and the costs resulting from imperfect energy conversion and storage. We also take into account the cost of replacing parts of the UPS over time under different conditions. We do not include the cost of installing the UPS. Because the HTS FESS is still under development, we use the performance of existing systems as a benchmark. When discussing the shortcomings of the existing FESS, we point out how the HTS FESS may achieve better performance, making a FESS an attractive substitute for a BESS in certain situations.

Other applications of energy storage systems include interim power for back-up system initiation and utility load-leveling. The decision to use batteries or a FESS in such applications is a complex one that depends upon the expected operating lifetime of the system, as well as its acquisition cost and operating cost. In general, FESS acquisition costs are higher than BESS acquisition costs. However, operating costs vary because they depend on the conversion efficiency and mode of usage of the energy storage system. Under the conditions that FESS operating costs can compensate for the higher FESS acquisition costs, performance criteria for flywheel- and battery-storage systems suggest that, beyond a threshold, long operating lifetimes and high duty cycles favor a FESS.

Energy Use Comparison of Flywheel and Battery Energy Storage Systems

Figure 5.1 summarizes energy flows for a generic storage device. At the left of the figure are the input and output connectors for the device to receive or provide

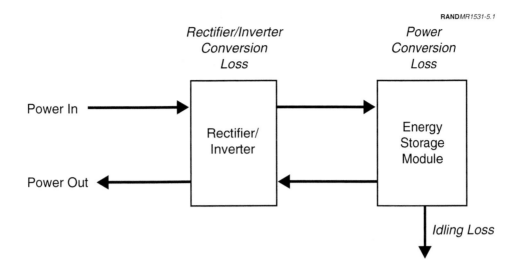

RAND*MR1531-5.1*

Figure 5.1—Schematic Representation of Power Flows in Energy Storage Systems

energy. The power conversion block—or rectifier/inverter—modifies the power from a fixed frequency (e.g., 60 Hz) AC to DC or a variable frequency AC and vice-versa so that energy can flow to and from the storage device. The operation of the power conversion block, as well as that of the energy storage module, results in losses that are noted at the top of the figure. The idling loss of the energy storage device, noted in the lower right corner of the figure, is defined as the energy drawn by the system when not in use.

The idling loss of an existing FESS, which requires continuous power input to keep the system fully "charged," is 1–3 percent of the power that the FESS can supply,[3] whereas the idling loss for a BESS, the battery self-discharge rate, is typically very small (e.g., 0.05 percent).[4] The rectifier/inverter conversion loss of a FESS and a BESS are similar because they employ similar electronic components and circuits. However, the power conversion losses differ because of differences in electromechanical and electrochemical energy conversion. Power conversion losses of a FESS are reported to be near 10 percent of the power handled whereas those of a BESS are reported to be in the 30–40 percent range.[5] These numbers suggest that a FESS will be a more attractive option when

---

[3]Wolsky, A. M., *The Status of Progress Toward Flywheel Energy Storage Systems Incorporating High-Temperature Superconductors*, Argonne National Laboratory, Argonne, Ill., October 17, 2000, pp. IV 8–9.

[4]Buchmann, Isidor, "Choosing a Battery That Will Last," *PowerPulse.Net*, http://www.powerpulse.net/powerpulse/archive/aa_040201b1.stm (last accessed August 31, 2001).

[5] Electromechanical Batteries," *Technology Profiles*, http://www.llnl.gov/IPandC/op96/12/12c-ele.html (last accessed August 23, 2000).

66

compared with a BESS at high utilization levels, for which idling losses will be less important than power conversion losses.

The time needed to recharge the storage device after use is also a key characteristic of a FESS or BESS. In FESS and BESS devices, the amount of time needed to return a set amount of energy to the energy storage module typically exceeds the amount of time needed to discharge this energy. Brochures on existing FESSs show that they take anywhere from 2.5 to 96 times longer to recharge than to discharge whereas two UPS systems that use batteries take 15 to 50 times longer to recharge than to discharge.[6] Because each FESS and BESS has different charging characteristics, detailed comparisons require specific system information. Later in this section, we present some general comparisons based on a parametric analysis of generic FESS and BESS characteristics. To normalize the comparisons, we analyze FESS and BESS systems with the same maximum energy storage and power output.

In Appendix C, we derive a parametric relationship between the storage system utilization and the losses per unit of power output for the FESS and BESS to have equal energy usage. The curved lines in Figure 5.2 illustrate this relationship and thus define the boundary between values of system utilization and FESS power conversion efficiency (the ratio of energy-storage module output to input power) such that the FESS uses less energy than a BESS with the same maximum energy storage and power output. Each curved line in the figure represents a different FESS idling loss ranging from 1–10 percent of output power. A FESS with values that appear above the curved lines in the figure would use less energy than an equivalent BESS. In plotting Figure 5.2, the BESS power-conversion loss was set at 30 percent of power output and the rectifier/inverter conversion loss was taken as 10 percent of power output for both systems. We note that system utilization has a different meaning here than it did in the cable and conductor case described in Chapter 4. Here, utilization is defined as the fraction of the time that energy is being drawn from the storage system. Thus, any system will have a maximum utilization that depends upon the ratio of its discharge time to its recharge time.

---

[6]See product description catalogs for UPS systems such as http://www.tic.toshiba.com/ups/files/1400sere44470secure.pdf and http://www.controlledpwr.com/products/md/md.pdf (last accessed August 24, 2001).

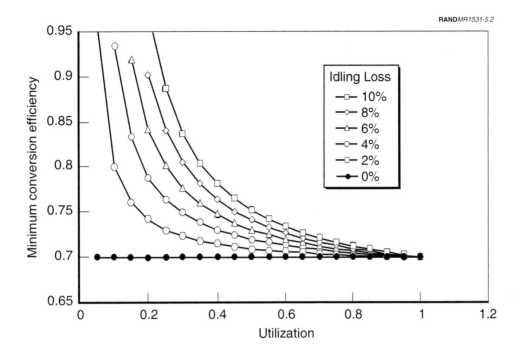

Figure 5.2—Energy Conversion Efficiency Required for a FESS to Compete with a BESS, Expressed as a Function of Utilization with Idling Loss Equal to 1–10 Percent of Output Power

As shown in Figure 5.2, the conversion efficiency needed for a FESS to compete with a BESS increases with idling loss. In addition, for any given energy-conversion efficiency, the FESS may have a minimum utilization rate below which it cannot compete with a BESS. For example, when the idling loss is 10 percent, the FESS utilization rate must be at least 25 percent if the energy-conversion efficiency is 94 percent. In fact, even if the energy conversion efficiency of the FESS is 100 percent, the utilization must still be at least 21 percent to compensate for idling losses. As noted earlier, FESSs are the most attractive option for applications in which they will be used at a high utilization rate. Also, any decrease in idling loss will translate into a proportionate improvement in performance for the FESS.

Compared with a BESS, a FESS begins to provide some advantages when it can recharge rapidly because this feature will increase its maximum utilization. One commercial FESS, manufactured by Beacon Power,[7] can recharge rapidly enough to meet the requirements for competing with a BESS. This FESS can operate for two hours and requires two hours or less to recharge, reaching a duty cycle greater than 50 percent. Based on the loss characteristics of this FESS, a duty cycle

---

[7]See www.beaconpower.com/products/BHE6_specs.pdf.

of 50 percent exceeds the 15 percent lower utilization limit for it to be more energy efficient than a BESS. In addition, this FESS recharges faster than a BESS, so it can operate in a regime that a battery cannot. Of course, the application must require a high utilization level in order to achieve these benefits. Utility load-leveling, in which the FESS could be charged when the cost of electricity is low and discharged when the cost of electricity is high in a repetitive periodic fashion (e.g., daily), would meet this requirement.

The recharge-to-discharge time ratios of commercial FESSs range from 1 to 96, providing duty cycles that range from 1 percent to 50 percent. The short duty cycle devices can be used for power quality management, but because of their idling losses (up to 10 percent of power output) these devices cannot compete with an equivalent BESS. Based on idling losses, discharge, and recharge times, the only FESS in existence today that can compete with a BESS is the one from Beacon Power.

Information on the power, expected duration of operation, and weight of FESSs reveals their intended applications. Using the power and duration of operation, we estimated the energy storage capability of each FESS and then calculated power and energy densities in W/kg and Wh/kg, respectively. These data are summarized in Table 5.1. When the manufacturers offer more than one model, the table presents illustrative values for simplicity. The parameters provided for the Boeing FESS, which uses HTS bearings, are based on development goals, because it is not yet commercially available. The parameters provided for the lead acid battery are representative and do not refer to a particular product.[8]

Table 5.1 clearly indicates that most of the FESSs were designed for power quality management rather than load-leveling. Their duration allows them to compensate for sudden short-term decreases in power. Except for the FESS offered by Beacon Power, these systems were not designed to operate for more than a fraction of a minute. This is reflected in the relatively high power densities and low energy densities of most of the FESSs as compared with the energy and power densities of a lead acid battery. Table 5.1 highlights the importance of matching energy and power performance when comparing a BESS and a FESS.[9]

---

[8]See http://ev.inel.gov/fop/general_info/battery.html.

[9]The FESS specifications used in this analysis reflect those of product offerings as of October 2001. Changes in FESS specifications of new product offerings may alter our detailed results slightly but will not change our conclusions.

Table 5.1

Comparison of FESS Model and Lead Acid Battery Characteristics

Manufacturer	Power (kW)	Duration (seconds)	Usable Energy (kWh)	Power Density (W/kg)	Energy Density (Wh/kg)
Beacon Power[a]	1	7,200	2	2.8	5.5
Acumentrics[b]	140	10	0.4	340	1
Active Power[c]	240	12.5	0.8	190	0.7
Trinity Flywheel[d]	100	15	0.04	730	3
Boeing[e]	100	300	8.3	670	56
Lead Acid Battery[f]	—	—	—	150	35

[a]See http://www.beaconpower.com/products/specs/specs.htm.

[b]See http://www.acumentrics.com/powerqueue-overview.html.

[c]See http://www.activepower.com/products/index.html.

[d]See http://www.afstrinity.com/specs.html.

[e]See Strasik, Michael, *Flywheel Electricity Systems with Superconducting Bearings for Utility Applications*, Annual Peer Review 2000, July 17–19, 2000, U.S. Department of Energy Superconductivity Program for Electric Systems. Proceedings on CD available from Energetics Inc., Columbia, Md., order number DE00759424.

[f]See http://ev.inel.gov/fop/general_info/battery.html.

The design proposed by Boeing for a FESS using HTS bearings anticipates higher power and energy densities than those that would occur with use of a lead acid battery. The use of HTS bearings and a low-loss electric generator are expected to improve the performance of Boeing's FESS so that it will achieve this type of performance.

Boeing's FESS is still under development. Even so, its performance can be estimated based on technical reports of losses. The bearing losses are expected to be on the order of 10 to 20 W for a 10 kWh FESS supplying either 3 kW or 100 kW, depending on the manner of its use.[10] Taking into account losses in the electrical part of the system provides an estimate of idling loss of less than 30 W for a FESS supplying either 3 kW or 100 kW, corresponding to between less than 0.03 percent and less than 1 percent. The major contributions to the electrical losses are from eddy currents, and these scale with the size of the generator,[11] so 1 percent is likely an overestimate. Using 1 percent as an upper limit, the FESS can easily compete against a BESS, based on the plotted lines shown in Figure 5.2. Assuming a conversion efficiency in the vicinity of 90 percent, similar to that of an existing FESS, this means that the Boeing FESS would still be an attractive option with a utilization as low as 5 percent. Because its recharge time is almost

[10]Strasik, Michael, *Flywheel Electricity Systems with Superconducting Bearings for Utility Applications*, Annual Peer Review 2000, July 17–19, 2000, U.S. Department of Energy Superconductivity Program for Electric Systems.

[11]Wolsky, A. M., *The Status of Progress Toward Flywheel Energy Storage Systems Incorporating High-Temperature Superconductors*, Argonne National Laboratory, Argonne, Ill., October 17, 2000.

70

equal to its discharge time, the Boeing HTS FESS should be able to operate at a utilization as high as 50 percent. This value exceeds the estimated minimum utilization that is required to compete with a BESS (see Figure 5.2) by one order of magnitude.

Life-Cycle Cost Estimates for a FESS and a BESS

Having established that a FESS may compete with an equivalent BESS based on energy use, we compare estimates of the life-cycle costs. Flywheels are designed to last 25 years. By comparison, batteries need to be replaced at relatively frequent intervals. Five years for batteries is an interval cited by manufacturers[12] and is used for our initial estimate. For purposes of the comparison, we use acquisition costs for an existing FESS manufactured by Beacon Power,[13] which are consistent with the cost of flywheel rotors and inverters. The battery cost cited by Beacon is $500, which is within the cost range of commercial systems. The cost of the inverter cited by Beacon is $1,300, which is at the low end of the range for commercial systems. The cost of commercial batteries is in the range of 23¢ per Wh to 33¢ per Wh, whereas the cost of an appropriate inverter is in the range of $1.47 per W to $2.77 per W based on the cost of a commercial battery back-up system.[14] When comparing FESS and BESS costs, however, the cost of the inverter is irrelevant because it is assumed that the FESS and BESS have the same type of inverter; therefore, the cost contribution of the inverter cancels out.

We use the same discount rate that we used in the cable and conductor comparisons in Chapter 4, 7 percent, and (the lower) electricity cost of 5¢ per kWh. Operating costs are computed for the 3 percent idling loss and 10 percent power conversion loss of the FESS, the 30 percent power conversion loss of the BESS, and the 10 percent rectifier/inverter conversion loss, all per unit of output power. Table 5.2 summarizes those costs. Utilization is 25 percent in this example. Installation and maintenance costs are not considered because they will vary according to location and usage. Table 5.2 shows that, with a 5-year replacement schedule for batteries, it takes between 10 and 15 years for the lower operating costs to pay back the higher acquisition cost of this particular FESS.

[12]Strasik, Michael, *Flywheel Electricity Systems with Superconducting Bearings for Utility Applications*, Annual Peer Review 2000, July 17–19, 2000, U.S. Department of Energy Superconductivity Program for Electric Systems.

[13]Hockney, Richard L., and Craig A. Driscoll, *Powering of Standby Power Supplies Using Flywheel Energy Storage*, Beacon Power Corporation, Woburn, Mass., http://www.beaconpower.com/company/assets/pdfs/paper_standby.pdf, (last accessed March 28, 2002).

[14]See, for example, the cost of the back-up systems from Thermo Technologies, Columbia, Md., on http://209.116.252.72/backup.htm (last accessed March 28, 2002).

Table 5.2

Acquisition and Life-Cycle Costs for a FESS and a BESS

	FESS	BESS
Acquisition cost	$3,650	$1,800
5-year life-cycle cost	$3,860	$2,090
10-year life-cycle cost	$4,010	$3,580
15-year life-cycle cost	$4,120	$4,640

The amount of time needed to recover the higher acquisition cost of the FESS depends on the replacement schedule of the BESS batteries. A more aggressive battery replacement schedule will result in a faster payback. Figure 5.3 shows the life-cycle costs for a 30-year lifetime FESS and BESS with various battery replacement schedules, based on a 7 percent discount rate. The life-cycle cost is shown as a function of time. For each year of operation, the acquisition cost and the discounted cost of operating for that number of years is shown. The FESS is never replaced. For the BESS, replacement schedules ranging from one to five

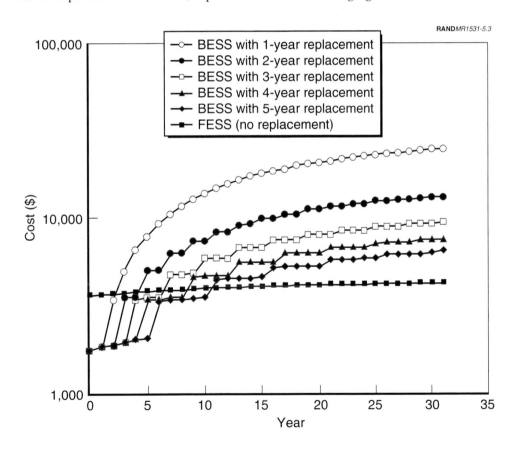

Figure 5.3—Effects of Replacement Schedule on Life-Cycle Costs for BESSs and a FESS

years are shown. Where the BESS lines cross the FESS line, the FESS is more cost-effective from that point on.

Because acquisition and operating costs are not currently available for Boeing's HTS FESS, it is not possible at this time to estimate the life-cycle cost of the system and to compare it with the cost of a BESS. However, the expected low losses of Boeing's FESS suggest that it may compete favorably with a BESS. As is the case with the Beacon Power FESS, the option of using a FESS will be especially attractive in applications that call for the frequent replacement of batteries in a BESS.

HTS Transformers

In this section, we draw comparisons between HTS and conventional transformers. First, we discuss our basic analysis and its limitations, and then present a comparison of energy loss and life-cycle cost estimates for HTS and conventional transformers.

Description and Limitations of the Analysis

The overwhelming majority of transmission and distribution of electricity in the United States makes use of AC currents. This is because of the ease with which AC voltages can be increased and decreased by using transformers, making transformers critical components of the U.S. electric grid. Generation is typically at 25 kV. For transmission, the voltage is increased to more than 100 kV at a lower current, which reduces conduction losses. In distribution networks, the voltage is gradually stepped down so that the current flowing in the lines outside a residential customer's home is at 240 V. Transformers are used to effect all of these voltage changes.

The following section describes energy loss and life-cycle cost comparisons of HTS and conventional transformers. These comparisons are based upon existing data on the engineering and cost characteristics of the conventional transformers and data and engineering estimates for HTS transformers based upon past experience with systems under development. We do not include the cost of siting and installing the transformers.

Energy Loss Comparison of HTS and Conventional Transformers

Electrical losses in existing transformers are very low. Efficiencies have been reported to reach 99.7 percent.[15] However, transformers are so pervasive throughout the electrical transmission and distribution network that the sum of these losses is significant, estimated at over one billion dollars annually with an electricity cost of 5¢ per kWh.[16] HTS transformers are expected to provide greater efficiencies than conventional transformers, but energy use for cooling reduces the expected energy savings. The impact of using HTS transformers is expected to depend upon their size because losses tend to scale nonlinearly with power ratings. In this analysis, we consider a 5 MVA HTS transformer, similar to one currently under development.[17]

In HTS and conventional transformers, two types of losses exist: conduction losses and core losses. Conduction losses are sometimes referred to as load losses, while core losses are sometimes referred to as no-load losses. In addition to these losses, HTS transformers also have thermal losses because they are operated at cryogenic temperatures and there is heat leakage through their electrical and thermal insulation to the ambient temperature external environment. The HTS transformer core is typically not cooled to a cryogenic temperature. Thus, the heat load that must be removed by the cryocooler does not include core losses.

We estimate core losses based on existing HTS and conventional transformer designs and data. Reported core losses for a 500 kilovolt amperes (kVA) HTS transformer built in Japan in 1996 were 2,300 W or 0.46 percent of the power capacity of the transformer.[18] This number is on the high end of open circuit losses in conventional transformers, which range from 0.2–0.5 percent of nominal power, with higher efficiency for larger transformers.[19] For example, an existing conventional 500 kVA transformer was reported to have core losses of 1,100 W.[20]

---

[15]Ramanan, V. R., Gilbert N. Riley, Jr., Lawrence J. Masur, and Steinar J. Dale, "A Vision for Applications of HTS Transformers," *International Wire & Cable Symposium Proceedings 1998*, p. 360.

[16]Mehta, Sam, as quoted in "High-Temp Superconductors Breed New Generation of Big Electric Equipment," *IEEE Spectrum*, January 2002, p. 22.

[17]"World's Most Powerful Superconducting Transformer (HTS) Successfully Tested," Waukesha Electric press release, May 27, 1998, http://www.waukeshaelectric.com/weshtspr.htm (last accessed August 31, 2000). See also Mehta, Sam, "5/10 MVA HTS Transformer SPI Project Status," *2001 Annual Peer Review Conference* (SUP), DOE/EE-SESAPR/2001-CD, available from the U.S. Department of Energy Office of Power Technologies, Washington, D.C.

[18]*HTS Transformer Development in Japan*, September 1997, http://itri.loyola.edu/scpa/03-07.htm (last accessed September 10, 2001).

[19]Chittawar, Milind, and G. Hari Kumar, *Transformers—Their Losses and Reduction*, http://www.letsconserve.com/paperenergy.htm (last accessed September 10, 2001).

[20]Technical Data Distribution Transformers: TUNORMA and TUMETIC, http://www.ev.siemens.de/download/pegt99/05_013.pdf (last accessed October 9, 2001).

74

Core losses range from 1,100 W to 1,700 W for a 1 MVA transformer, whereas the estimated core loss in a HTS 1 MVA transformer is 2,300 W.[21] Thus, using current designs as a basis, the HTS transformer core losses appear to be somewhat greater than those of conventional transformers. Using existing transformers as a baseline for core losses, a 5 MVA transformer's core loss is expected to be between 7,000 and 14,000 W.

We estimate conduction losses from data on losses in the wire used to wind the transformer coils. For copper cables used for electricity transmission, losses range from 15 watts per kiloampere meter (W/kAm) to 40 W/kAm.[22] Other estimates suggest losses in copper cables as high as 80 W/kAm.[23] By comparison, the AC losses in the BSCCO wire that is wound to form the coils of HTS transformers were estimated to be as low as 0.25 W/kAm.[24] This figure is consistent with the previously reported estimate of AC losses of 1 W/m per phase in HTS cables with a capacity of 3,000 A.[25] However, this goal was not achieved for the multikilometer lengths of wire required for a demonstration transformer.[26]

Waukesha Electric's 5 MVA HTS transformer design calls for 10 km to 20 km of wire per phase. If the current being carried is approximately 100 A, and the wire-loss goal described in the previous paragraph is achieved for production of the 10 km to 20 km of wire required, the wire loss would be 250 W to 500 W. This number is consistent with an independent estimate that a 10 MVA HTS transformer would dissipate approximately 900 W from conduction losses.[27]

We based the estimates for the thermal losses from heat leakage through the insulation and terminations on data from HTS transformer demonstrations. The loss from heat leakage of a demonstration 500 kVA HTS transformer built in

[21]Schwenterly, S. W., et al., *IEEE Transactions on Applied Superconductivity*, Vol. 9, No. 2, June 1999, p. 680.

[22] *Underground Transmission Cable High Voltage (HV) and Extra High Voltage (EHV) Crosslinked Polyethylene (XLPE) Insulation Extruded Lead or Corrugated Seamless aluminum (CSA) Sheath 69-345 kV, Copper or Aluminum Conductor*, BICC Cables Company, West Nyack, N.Y.

[23]Marsh, G. E. and A. M. Wolsky, *AC Losses in High-Temperature Superconductors and the Importance of These Losses to the Future Use of HTS in the Power Sector*, Argonne National Laboratory, Argonne, Ill., May 18, 2000.

[24]Ramanan, V. R., Gilbert N. Riley, Jr., Lawrence J. Masur, and Steinar J. Dale, "A Vision for Applications of HTS Transformers," *International Wire & Cable Symposium Proceedings 1998*, p. 360.

[25]Nassi, Marco, Pierluigi Ladie, Paola Caracino, Sergio Dreafico, Giorgio Tontini, Michel Coevoet, Pierre Manuel, Michele Dhaussy, Claudio Serracane, Sergio Zannella, and Luciano Martini, "Cold Dielectric (CD) High-Temperature HTS Cable Systems: Design, Development and Evaluation of the Effects on Power Systems," manuscript describing the Pirelli cable demonstration effort, received by RAND September 19, 2000.

[26]Zueger, Harry, *Transformateur supraconducteur à haute température 10 MVA*, Swiss Federal Office of Energy, September 2000.

[27]Marsh, G. E. and A. M. Wolsky, *AC Losses in High-Temperature Superconductors and the Importance of These Losses to the Future Use of HTS in the Power Sector*, Argonne National Laboratory, Argonne, Ill., May 18, 2000.

Japan was measured to be 51 W, including losses through the electrical leads.[28] Thus, we use 500 W as a conservative estimate of the heat leakage for the 5 MVA transformer.

The energy use of the cryocooler depends upon temperature and cryocooler efficiency. The expected operating temperature of an HTS transformer varies with design. Waukesha Electric has tested a 1 MVA demonstration transformer that operated in the vicinity of 25 K.[29] ABB designed a 10 MVA demonstration transformer with an operating temperature of 68 K.[30] The difference in operating temperatures of these two HTS transformers leads to a significant difference in estimated cryocooler energy use. For example, assuming that the cryocooler operates at 20 percent of its maximum possible (Carnot) efficiency,[31] the cryocooler for the 68 K transformer requires 16.9 W of input power to remove each watt of heat load at the operating temperature. For the 25 K transformer, a cryocooler operating at 20 percent Carnot requires 54.6 W of input power for each watt of heat load. Accordingly, the total loss of the HTS transformer ranges from 20 kW to 69 kW, or 0.4 percent to 1.4 percent, of the rated power, respectively. By contrast the total loss of a 5 MVA conventional transformer is about 40 kW, or 0.8 percent of the rated power.[32] Thus, the operating temperature of the HTS transformer strongly influences the energy use comparison.

As noted earlier, the HTS transformer loss also depends upon the amount of wire used in its windings because the wire loss is given in W/m. Appendix D derives an expression for the loss of the HTS and conventional transformers based upon the loss terms described earlier. Figure 5.4 shows the dependence of the total loss of a 5 MVA HTS transformer on operating temperatures for two different wire lengths based upon the expression in Appendix D, using the parameters estimated earlier and assuming a cryocooler efficiency of 20 percent Carnot.

---

[28]*HTS Transformer Development in Japan*, September 1997, http://itri.loyola.edu/scpa/03-07.htm (last accessed September 10, 2001).

[29]Schwenterly, S. W., et al., *IEEE Transactions on Applied Superconductivity*, Vol. 9, No. 2, June 1999.

[30]Zueger, Harry, *Transformateur supraconducteur à haute température 10 MVA*, Swiss Federal Office of Energy, September 2000.

[31]See the discussion of cryocooler efficiency in Chapter 4.

[32] *Technical Data Distribution Transformers: TUNORMA and TUMETIC*, http://www.ev.siemens.de/download/pegt99/05_013.pdf, (last accessed March 29, 2002).

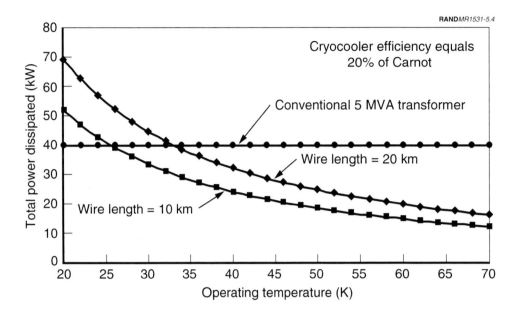

Figure 5.4—Dependence on the Operating Temperature of the Total Power Dissipated by a 5 MVA HTS Transformer

Figure 5.4 demonstrates that a 5 MVA HTS transformer with 20 km of wire operating at 33 K or with 10 km of wire operating at 26 K has a loss equal to that of a conventional 5 MVA transformer. Thus, for operating temperatures above 33 K or 26 K, depending upon the length of the wire, the HTS transformer with a 20 percent Carnot cryocooler will dissipate less power during operation than will a conventional transformer. The Waukesha Electric 5/10 MVA demonstration transformer anticipates "closed-cycle cryocooling of the transformer windings to temperatures as low as 382 degrees below zero, Fahrenheit (F),"[33] or 43 K.

Appendix D derives a parametric relationship between the cryocooler efficiency and the transformer utilization for the HTS transformer to dissipate less energy than the conventional transformer. Here, utilization is defined as in Chapter 4 as the fraction of its maximum rating at which the transformer is operated, and the same approximation to the load curve is used, resulting in a linear dependence of the conduction losses on utilization. As for the cable and conductor comparisons, this approximation will be best for high utilization (e.g., $u > 0.6$). Figure 5.5 shows curves that satisfy this parametric relationship and thus define the boundary between values of cryocooler efficiency and utilization for which the HTS transformer uses less energy than the conventional transformer (the region above the curved lines in the figure) and for which the HTS transformer uses

---

[33] "World's Most Powerful Superconducting Transformer (HTS) Successfully Tested," http://www.waukeshaelectric.com/weshtspr.htm (last accessed August 31, 2000).

more energy than the conventional transformer (the region below the curved lines).

Figure 5.5 uses loss parameters for a 5 MVA HTS transformer with a wire length between 10 km and 20 km, and shows operating temperatures of 43 K and 68 K corresponding to the demonstration transformers discussed earlier. In both cases, high values of utilization and cryocooler efficiencies near 20 percent are required to produce HTS transformer energy savings. However, we note that energy savings are not the sole or most important benefit of HTS transformers. Other important benefits of HTS transformers that have been demonstrated include weight and size reduction for the same maximum power rating and fault current protection.[34]

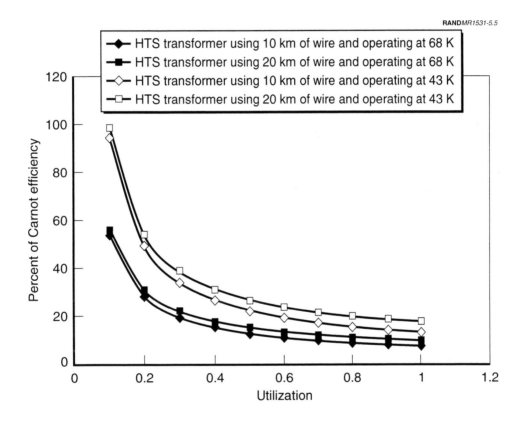

Figure 5.5—Boundaries Showing that a 5 MVA HTS Transformer with Cryocooler Efficiency and Utilization Values Above the Curves Uses Less Energy than a Conventional Transformer

---

[34]Schwenterly, S. W., S. P. Mehta, and M. S. Walker, "HTS Power Transformers," in 2001 Annual Peer Review Conference (SUP), DOE/EE-SESAPR/2001-CD, available from the U.S. Department of Energy Office of Power Technologies, Washington, D.C.

78

Life-Cycle Cost Estimates for HTS and Conventional Transformers

Figure 5.5 shows that, under appropriate circumstances, an HTS transformer can provide energy savings compared with a conventional transformer. We now compare the life-cycle cost of the two systems. As for the analysis of HTS cables, we assume a 7 percent discount rate. It is expected that the HTS transformer will cost more to acquire than a conventional transformer because it requires a cryocooler in addition to typical transformer components. Thus, for the HTS transformer to provide a life-cycle cost benefit, the savings resulting from its more efficient operation (which will depend upon the cost of electricity) must be sufficient to overcome its higher acquisition cost. Appendix D presents an expression for the life-cycle cost of HTS and conventional transformers in terms of their acquisition costs and discounted yearly operating costs, from which a parametric relationship between the HTS acquisition cost and cryocooler efficiency for the HTS transformer to provide life-cycle cost savings can be derived.

Figures 5.6 and 5.7 show the discounted payback for HTS transformers, using 10 km of wire and operating at 68 K, derived from this relationship with a cryocooler cost of $100 per W of heat load (the value used earlier for a low-volume commercial cryocooler) and the HTS wire cost (the predominant cost element) used as a proxy for the acquisition cost. The acquisition cost of a conventional 5 MVA transformer is in the range of $50,000 to $75,000;[35] the Figures 5.6 and 5.7 use the midpoint of this range, $65,000. As in the HTS cable section, we assume electricity cost in the range of 5¢–10¢/kWh.

Figures 5.6 and 5.7 also show that for the assumed parameters (e.g., $100 per W cryocooler cost, 10 km wire requirement) HTS transformers will not compete favorably with conventional transformers at the current HTS wire cost of $200 per kAm. However, at the projected 2004 HTS wire cost of $50 per kAm,[36] Figures 5.6 and 5.7 suggest that a five- to ten-year discounted payback should be possible for cryocooler efficiencies that are currently achievable. We note, however, that the 10 km of HTS wire used in the transformer must also meet the performance requirement of 0.25 W/kAm conduction loss to achieve this result.

In closing this section, we note that, as with HTS cables, energy savings and life-cycle cost benefits are not likely to be the driving factors in a decision to use HTS transformers. Weight and size reduction for the same maximum power and the

---

[35]Mehta, S., private communications, Waukesha Electric, 2001.

[36]Howe, John, American Superconductor, briefing on superconductivity at Rayburn House Office Building, August 3, 2001.

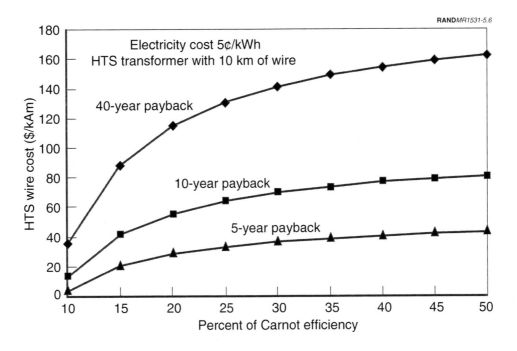

Figure 5.6—Discounted Payback Time for a 5 MVA HTS Transformer Compared with a
Conventional Transformer as a Function of HTS Wire Cost and Cryocooler Efficiency,
Electricity Cost of 5¢ per kWh

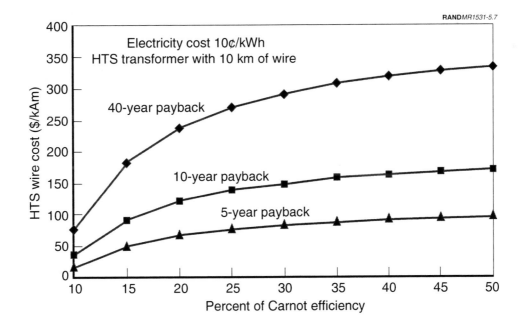

Figure 5.7—Discounted Payback Time for a 5 MVA HTS Transformer Compared with a
Conventional Transformer as a Function of HTS Wire Cost and Cryocooler Efficiency,
Electricity Cost of 10¢ per kWh

environmental and safety benefits of operation at cryogenic temperatures with liquid nitrogen (rather than flammable oil) as a coolant, all of which provide siting flexibility, may well be more important factors in the decision to use an HTS rather than a conventional transformer.

Superconducting Magnetic Energy Storage

A superconducting magnetic energy storage system is composed principally of a coil of superconducting wire and accompanying power electronic components. The SMES stores energy in a direct current flowing through the superconducting coil. When drawing energy from a SMES device, a power electronics module changes the DC current flowing through the SMES into an AC current that matches application needs. The same power electronics module changes the external AC current into a DC current that can flow through the SMES coil. The response time of the power electronics module is less than half a millisecond and the SMES can supply current for several seconds.[37] With this rapid response, SMES can be used either to provide power during brief outages for an electricity user for whom power is critical, such as an industrial plant, or as a tool for power regulation on the grid.

The existing commercial SMES device uses both HTS and LTS materials. The coil through which the current flows is made from LTS NbTi wire and is cooled with liquid helium. The leads connecting the coil to the power electronic components at room temperature are made of HTS BSCCO. Evaporating helium inside the SMES cools the leads. This arrangement is more energy efficient than earlier designs that connected the LTS material directly to normal metal leads.[38]

Comparing SMES and FESS Performance

The energy stored in a SMES is comparable to that of most commercial FESSs (see Table 5.1). The energy storage of the micro-SMES offered by American Superconductor is on the order of 3 MJ, which equals 0.83 kWh.[39] However, the major difference between SMES and FESS is in the power delivered. With the exception of the Beacon Power system, FESSs are used to deliver hundreds of kilowatts of power. By contrast, SMES delivers megawatts of power. American

---

[37]American Superconductor, *D-SMES Data Sheet*, http://www.amsuper.com.

[38]Schwall, Robert, American Superconductor, presentation at the Applied Superconductivity Conference 2000, Virginia Beach, Va., September 18, 2000.

[39]Note that the energy storage of the Beacon Power FESS is 2 kWh, and the energy storage of the Boeing HTS FESS is 8.3 kWh.

Superconductor offers a number of different micro-SMES products with different applications. For example, its D-SMES product used by Wisconsin Public Service's Northern Transmission Grid can provide 8 megavolt amperes of continuous reactive power[40] (8 MVAR) with a peak of 20 MVAR. The magnet has a peak discharge of 3 MW. Another D-SMES product is rated for a continuous output of 16 MVAR and peak of 28 MVAR. PQ-IVR is a SMES system for industrial customers and is rated at a peak discharge power of 3 MW of real power and a peak of 17 MVAR of instantaneous reactive power. In this case, the magnet stores 2.7 MJ of energy. [41]

The difference in performance between a SMES and a FESS lies primarily in their power electronics modules that constitute the inverter. The inverter in SMES is designed to handle about ten times the power of the inverter in FESS and can be used to change the phase between an input voltage and output voltage. In this way, the inverter can generate a voltage that lags or leads the voltage on the grid. If this is done without supplying or removing power from the grid, the effect is a phase shift between the current and the voltage.[42] Thus, the power electronics module in SMES can be used to generate reactive power.

The output of the inverter in SMES is not perfectly sinusoidal. For this reason, the instantaneous input and output power of the inverter may not be exactly equal, and a DC source may be necessary to make up the shortfall or absorb the excess. The SMES coil can be used in such a case. In addition, the SMES coil can be used when power is needed for short periods of time and sufficient power is not available from the grid.[43] When equipped with a DC bus, SMES can supply reactive power continuously, a feature of any system that combines an inverter with an appropriate storage medium.

Potential Uses of SMES

The growth of computer systems in recent years has put a spotlight on the need for clean power. Computer controllers have little tolerance for excursions from the desired power input. As noted in the "Flywheel Energy Storage Systems" section, the effects of excursions away from the rated voltage have been compiled

[40]See Footnote 4 in Chapter 3 and Taylor, Carson W., "Improving Grid Behavior," *IEEE Spectrum,* June 1999, pp. 40–45.

[41]Schwall, Robert, American Superconductor, presentation at the Applied Superconductivity Conference 2000, Virginia Beach, Va., September 18, 2000. See also http://www.amsuper.com.

[42]Gyugyi, Laszlo, "Solid-State Synchronous Voltage Sources for Dynamic Compensation and Real-Time Control of AC Transmission Lines," *Emerging Practices in Technology,* Institute of Electrical and Electronics Engineers, Inc., 1993.

[43]Gyugyi, Laszlo (1993).

82

in the CBEMA Curve.[44] This curve shows that effects on equipment depend on duration and the extent of deviation from the desired voltage. For example, electronic equipment can tolerate 300 percent of nominal voltage for no more than 100 µs. This value drops to 200 percent of nominal voltage for 1 ms and 106 percent of nominal voltage for times exceeding tens of seconds. Generally, power glitches that last a few milliseconds are not logged as failures by the electric utilities because their effects are minor or even imperceptible for many customers.[45] Thus, disruptions to electric distribution that may pass unnoticed in many applications become critical for computer systems.

The response of a SMES system requires less than 0.5 ms from the time the fault is detected.[46] This makes it well-suited to improve power quality when a rapid response is needed. Areas where SMES systems have been tested and installed include manufacturing facilities, military buildings, and a semiconductor testing facility. In those areas, the SMES systems were used primarily to protect motors.[47] Although these applications of SMES back-up systems provide useful illustrations of their capabilities, facilities dependent upon computer systems may provide better insights into the use of SMES for power conditioning and back-up when a fast response is needed.

With respect to grid support, Wisconsin Public Service Corporation has purchased six SMES units to enhance the stability of a portion of its power grid in the northern part of Wisconsin, which has experienced power sags as a result of growth in demand in recent years. The long-term solution to the growth in demand is the construction of a new transmission line. In the short term, however, the utility has decided to rely on SMES to meet bursts of demand.[48] The SMES will be used to boost voltage at the onset of disturbances along the system.

Whether or not SMES is a viable and cost-effective power quality component depends upon the details of the application. In the case of Wisconsin Public Service, cost estimates were made for six D-SMES systems, static volts amperes reactive (VAR) compensation, and additional transmission lines, and SMES was

---

[44]Dorrough Electronics Inc., *CBEMA Curve*, 1999. See http://www.dorrough.com/ What_s_New/PLM-120/PowerMonTM/CBEMA_Curve/cbema_curve.html (last accessed March 28, 2002).

[45]Walk, Matthew L., "Energy to Count On," *The New York Times*, August 17, 1999, pp. B1, B7.

[46]American Superconductor, *D-SMES Data Sheet*, http://www.amsuper.com.

[47] Gravely, Michael L., and Jeffrey Goldman, *Proceedings, Naval Symposium on Electric Machines*, Newport, R.I., July 28–31, 1997.

[48]Bogard, Larry, "Grid Voltage Support," *Transmission and Distribution World*, October 1999.

found to be the lowest-cost option.[49] At the quoted cost of $4.8 million for six systems rated at 8 MVAR continuous support, the D-SMES option cost $100,000 per MVAR. If the 20 MVAR of instantaneous support is used in the estimate, the cost drops to $40,000 per MVAR, which is comparable to the estimated cost of $30,000 to $40,000 per MVAR for conventional VAR compensation.[50]

The ability of distributed SMES systems to store electricity at desired locations and to respond to voltage sags and other power quality problems within a small fraction of a cycle allows the transmission and distribution network to operate under conditions of high load within a tighter margin, which allowed the Wisconsin Public Service to postpone new transmission line additions. The fact that the SMES systems are distributed and transportable (trailer-mounted) allows the flexibility to modularize the investment and to reallocate the voltage support resources wherever they are needed as the transmission and distribution network and the load it serves grow and evolve over time. This may provide a niche market for distributed SMES even if its per-MVAR cost is higher than that of conventional VAR compensation. While SMES systems cannot substitute for new transmission lines, they can allow the existing system to be operated more effectively by providing voltage support to prevent voltage collapse under conditions of heavy loads and long-distance power transfers.

Three projects that will place additional D-SMES systems on utility grids have been announced at the time of this writing: Alliant Energy has ordered one system for their Wisconsin grid,[51] Entergy has ordered four systems for its Texas grid,[52] and TVA has ordered one system that was planned for deployment on its power grid in November 2001.[53] According to American Superconductor, the deployment of the TVA system brings the number of D-SMES systems installed since July 2000 on the U.S. utility grid to ten.

---

[49]Schwall, Robert, American Superconductor, presentation at the Applied Superconductivity Conference 2000, Virginia Beach, Va., September 18, 2000.

[50]Dale, Steinar, private communications, ABB, 2000.

[51]"Alliant Energy Places Order for D-SMES Unit," American Superconductor Press Release, March 30, 2000, http://www.amsuper.com

[52]"American Superconductor and GE Receive First SMES Order: Entergy to Utilize D-SMES to Assure Power Reliability," American Superconductor press release, September 30, 2000; and "American Superconductor and GE Industrial Systems Receive Follow-on D-SMES Order from Entergy: Entergy Chooses D-SMES to Assure Power Reliability in Summer of 2001 and 2002," American Superconductor press release, May 7, 2001, http://www.amsuper.com.

[53]"American Superconductor and GE Receive D-SMES Order from TVA: TVA Becomes Fourth Utility to Choose American Superconductor's Technology to Assure Power Reliability," American Superconductor press release, September 11, 2001, http://www.amsuper.com.

6. Conclusions and Federal Action Plan

In this final chapter, we present our conclusions regarding ways in which HTS power technologies may be used in the future to address the growing demands on the U.S. power transmission grid. We also review the recommendations from the OSTP's 1990 national action plan on superconductivity R&D that relate to HTS and suggest recommendations for a 2002 action plan.

Conclusions

Based upon our observations concerning the impact of the evolving electricity market on the U.S. transmission grid, the simulation results, and the engineering analyses described in this report, we draw the following conclusions:

1. Significant transmission constraints exist in many areas of the U.S. These constraints have resulted from increased demand, increased power transfers, and very small increases in transmission capacity over the past several years.

2. These transmission constraints have contributed in some cases to decreased reliability and price differentials between load areas.

3. High-temperature superconducting underground cables provide an attractive retrofit option for urban areas that have existing underground transmission circuits and wish to avoid the expense of new excavation to increase capacity. This situation exists because HTS cables have almost zero resistance, very small capacitance and inductance, and high power capacity compared with conventional cables of the same voltage. Thus, the HTS cables provide changes in power flows that reduce stress in heavily loaded circuits, thereby increasing reliability or power transfer capacity and relieving transmission constraints.

4. When operated at high utilization, HTS cables provide energy savings benefits compared with conventional cables per unit of power delivered for a range of HTS cable parameters consistent with existing data and engineering estimates. However, whether or not the concomitant HTS cable operating cost savings are greater than the increase in acquisition cost compared with conventional cables depends upon the cost of electricity.

5. HTS cables can provide a parallel transmission path at lower voltage to relieve high-voltage transmission constraints. The implementation of this approach for long-distance transmission circuits will depend on the

86

development of periodic cooling stations and sufficient manufacturing capacity for HTS wire.

6. When operated at high utilization, HTS cables provide energy savings benefits compared with conventional overhead lines per unit of power delivered for a range of HTS cable parameters consistent with existing data and engineering estimates. HTS cables may also provide concomitant life-cycle cost benefits for situations in which the usual cost advantage of overhead transmission lines is mitigated by site-specific concerns, such as high land use demands or right-of-way costs, or the expense of obtaining siting approvals or increasing power transfer capacity at higher voltage.

7. Flywheel energy storage systems using HTS magnetic bearings are in the demonstration stage and have the potential to achieve performance characteristics that will make flywheels competitive with batteries in a wide range of electricity storage applications.

8. HTS transformers can provide increased power capacity with the same footprint as conventional transformers and could be sited inside buildings because they eliminate fire hazards associated with oil. If estimated HTS wire cost reductions from a new manufacturing facility are achieved and the wire meets performance requirements, HTS transformers are projected to be cost competitive with conventional transformers for a range of parameters consistent with existing data and engineering estimates.

9. Superconducting magnetic energy storage systems that use low-temperature superconducting coils and HTS current leads have already found a niche market for distributed reactive power support to prevent grid voltage collapse and for maintaining power quality in manufacturing facilities.

Federal Action Plan for HTS Research and Development

During the early years of HTS research, the White House OSTP developed a national action plan on superconductivity R&D that included the following recommendations related to HTS:[1]

• Continuation of superconductivity R&D with an emphasis on HTS through a cooperative effort involving federal laboratories, industry, and universities, that is coordinated by OSTP

---

[1]*The National Action Plan on Superconductivity Research and Development Annual Report for 1990,* Executive Office of the President, Office of Science and Technology Policy, February 1991, p. 12.

- Strengthening of the program to develop HTS wire for commercial products, led by the DOE and the national laboratories

- Increasing attention paid to the development of the enabling technology of compact refrigeration to enhance the performance and reduce the cost of superconducting (and semiconducting) devices, led by the DoD and the National Aeronautics and Space Administration (NASA).

The implementation of these recommendations was mixed. The DOE High-Temperature Superconductivity for Electric Systems Program has led a decade-long effort that has made substantial progress on wire development and is now focusing on second-generation wire to reduce cost. Through the SPI, the DOE and a number of industrial concerns are jointly funding several HTS power technology demonstration projects. However, by emphasizing compact refrigeration and giving the lead to the DoD and NASA, the 1990 OSTP recommendation led to the development of small and expensive custom cryocoolers that are not appropriate for electric power systems.[2] Thus, the need still exists for development of affordable cryocoolers with multikilowatt capacity to provide cooling for HTS power system components, such as cables and transformers.

Based upon the analysis described in previous chapters, a 2002 action plan for HTS power technologies should include the following recommendations:

- The DOE-led HTS power technologies R&D program should continue to emphasize second-generation wire development, with the goal of providing HTS wire meeting commercial cost and performance targets.[3]

- The SPI should be expanded, while building on the current demonstrations that are providing operating experience, to develop new demonstrations with operational energy or power-transfer benefits. These new demonstrations should include HTS cable demonstrations at longer lengths and transmission voltages and demonstrations of HTS transformers and FESSs at a scale consistent with utility substation and end-user facility needs.

- The HTS R&D program should increase the emphasis on and support for the development of cryocoolers with multikilowatt capacity that can be mass produced. These cryocoolers should have an efficiency rating greater than

---

[2]Nisenoff, M., "Status of DoD and Commercial Cryogenic Refrigerators," in *Cryogenics Vision Workshop for High-Temperature Superconducting Electric Power Systems Proceedings*, U.S. Department of Energy Superconductivity Program for Electric Systems, July 1999.

[3]*Proceedings Coated Conductor Development Roadmapping Workshop—Charting Our Course*, U.S. Department of Energy Superconductivity for Electric Systems Program, January, 2001.

today's range of 14–20 percent of Carnot at the HTS-power-component operating temperature.

- The HTS R&D program should increase the emphasis on and the support for the development of standards for HTS-power-component installation and operation and increase the emphasis on and the support for training of industrial staff members who will operate and maintain these installations.

A. Technical Characteristics and Demonstration Status of HTS Power Technologies

This appendix supplements the discussion in the body of the report by supplying information on the technical characteristics and demonstration status of HTS cables, flywheel energy storage systems, and HTS transformers. This material is not intended to be comprehensive but rather is meant to provide an informational context for readers with less familiarity with these technologies.

HTS Cables

In this section, we briefly describe the technical characteristics of the two principal types of HTS cable designs and provide data on the status of HTS cable demonstrations worldwide.

Technical Characteristics

HTS cables use one of two designs: warm dielectric or cold dielectric;[1] the two designs offer distinct advantages that make one type of cable preferable over the other for particular applications.

The warm dielectric cable, which carries only one phase of current, is relatively simple in design. At its core is a duct through which the cryogenic medium flows. The duct is surrounded by the HTS wire, which is insulated with a dielectric sleeve. Over the dielectric is a protective sheath. The whole assembly fits into a pipe or duct.

The cold dielectric cable can be configured in different ways, depending on how many layers are concentric and depending on where the cryogen flows. One design calls for each phase to consist of a central cryogen duct surrounded by a support and the HTS wire. A dielectric separates the inner conductor that carries the current from an outer HTS sleeve that acts as a return and shields the inner

---

[1]"Selected High-Temperature Superconducting Electric Power Products," Superconductivity Program for Electric Systems, Office of Power Technologies, U.S. Department of Energy, January 2000.

90

conductor from external magnetic fields. Skid wires surround the return conductor. Because the three phases are adequately shielded from one another, they can fit together in a common cryostat where they are bathed in a cryogen bath.

In an alternative cold dielectric cable, the core of each phase consists of a cryogen surrounded by a support and the HTS wire. A dielectric separates the inner conductor from an outer HTS sleeve, around which a cryogen flows. Each phase of the cable in this assembly is contained by a cryostat encircled by skid wires. The three phases lie inside a common pipe or duct.

The main advantage of the warm dielectric cable is that its impedance matches that of a conventional cable of similar diameter. Such a warm dielectric cable can easily substitute for a conventional cable in a retrofit. This is the case with the Detroit Edison demonstration cited in Chapters 2 and 4. With the same impedance as the cable it replaces, the warm dielectric cable will not affect the flow of power through other cables. However, its higher current-carrying capacity can be used to relieve a bottleneck in a transmission grid. The simpler design of the warm dielectric cable compared with the cold dielectric cable also translates into a lower acquisition cost.

Because the warm dielectric cable lacks a conducting shield, it is sensitive to external magnetic fields. As a result, its losses will be higher than those of a cold dielectric cable. Also, the lack of shielding calls for the appropriate positioning of the three phases to minimize the magnetic fields they induce in one another. The cold dielectric cable is more robust because each phase is protected from the magnetic field of the other phases. However, its low impedance requires a careful analysis of the effects of a new cable on the entire circuit in which it will be installed.

Table A.1 compares the power capacity and losses of a conventional cable with HTS warm and cold dielectric cables that could substitute for the conventional cable in a retrofit using existing conduits.

Demonstration Status

HTS AC cables are currently under consideration and are in a demonstration status to replace conventional AC underground cables and overhead lines in situations in which additional capacity is needed and siting issues preclude conventional approaches. Several demonstrations and tests of HTS cables have been or are being performed in the United States and abroad, as summarized in Table A.2. Two tests of particular interest are listed in the last two rows of the

Table A.1

Collected Properties of Conventional, HTS Warm Dielectric, and HTS Cold Dielectric Cables

	Conventional	HTS Warm Dielectric	HTS Cold Dielectric
Pipe outer diameter (inches)	8	8	8
Voltage (kV)	115	115	115
Power (MVA)	220	500	1,000
Losses (W/MVA)	300	300	200

SOURCE: Nassi, Marco, Stephen Norman, Nathan Kelley, Christopher Wakefield, David Bogden, and Jon Jipping, "High Temperature Superconducting Cable System at Detroit Edison," presented at the 2000 T&D World Expo, April 26–28, 2000, Cincinnati.

Table A.2

HTS Cable Demonstration Projects

Participants	Voltage	Current	Length of Cable
EPRI, Pirelli, DOE	115 kV	2 kA	50 m
DOE, IGC, Southwire	12 kV	1.25 kA	5 m
Tokyo Electric Power Company, Sumitomo	66 kV	1 kA	30 m
Electricité de France, Pirelli	225 kV	3 kA	50 m
Pirelli, Siemens	110 kV	2 kA	50 m
NKT Cables, Nordic Superconductor Technologies A/S, Danmarks Tekniske Universitet, and other foreign companies and utilities	60 kV	2 kA	30 m
ENEL SpA, Pirelli, Edison SpA	132 kV	3 kA	30 m
EPRI, Pirelli, DOE, Detroit Edison	24 kV	2.4 kA	120 m
Southwire/DOE/ Oak Ridge National Laboratory	12.5 kV	1.25 kA	30 m

SOURCE: "Câble Supraconducteur," Electricité de France Presentation at *Sixièmes Journées du Froid: Commission Cryogénie et Supraconductivité*, May 16–19, 2000.

table. These tests are ongoing in the United States and are explained in more detail later in this section. In all cases, the cables are several meters to tens of meters long and the current flowing through them is in the range of 1,000 to 3,000 amperes. Tests of HTS cables are taking place in urban settings where demand for electricity strains the existing systems and building additional transmission capacity is impractical.

In Detroit, three warm dielectric HTS cables built by Pirelli using HTS material from American Superconductor, each carrying 2,400 A at 24 kV, are used to replace nine traditional copper lines in Detroit Edison's Frisbie substation.[2] The

[2]Superconductivity Partnership Initiative, *Power Cable (Retrofit) Project Fact Sheet*, revised March 16, 2000, U.S. Department of Energy, Washington, D.C.

existing infrastructure contains nine conduits, so this setup leaves six conduits that can in principle eventually be used for additional HTS cables, thus increasing the capacity of the system by a factor of three. (This level of capacity increase would require some redesign because, as noted earlier, the warm dielectric cable design would not allow such proximity due to overlapping magnetic fields.)

In another U.S. demonstration, Southwire has built a cold dielectric HTS cable to provide power to three manufacturing plants.[3] The HTS material was provided by Intermagnetics General Corporation (IGC). The cable differs somewhat from the one used in Detroit. In this case, a 30 m three-phase cable carries 1,250 A at a voltage of 12.5 kV. Power losses associated with this HTS cable are estimated at 0.6 W/kAm from AC losses and 1 W/m per phase for the cryostat. This is equivalent to three-quarters of 1 percent, compared with 5 to 8 percent for copper cables.[4] This cable can carry three to five times more current than a copper cable of similar size and has logged more than 10,000 hours at full capacity as of this writing.[5]

As a result of the success of the HTS cable demonstrations, the DOE recently announced new projects to install and test longer HTS cables in utility networks. One project is the installation and demonstration of a 77 MVA, three-phase, 2,500-foot long, HTS cable system in a congested urban area on Long Island. The project will be conducted by Pirelli Cables & Systems, American Superconductor, Keyspan—Long Island Power Authority, EPRI, and Air Liquide. The project will also develop a 1 kW (cold) pulse tube refrigerator for high-efficiency cooling.[6]

A second project is the demonstration of a 1,000-foot-long, three-phase HTS cable at a substation in Columbus, Ohio. The project team includes Southwire Company, American Electric Power, PHPK, Nordic Superconductor Technologies, 3M Company, Integrations Concepts Enterprises, and Oak Ridge National Laboratory. The goal of the project is to replace an existing oil-filled underground power cable that has limited current carrying capacity.[7] In a third

---

[3]Superconductivity Partnership Initiative, *Power Cable Project Fact Sheet*, revised July 11, 2000.

[4]Marsh, G. E. and A. M. Wolsky, "AC Losses in High-Temperature Superconductors and the Importance of These Losses to the Future Use of HTS in the Power Sector," Argonne National Laboratory, Argonne, Ill., May 18, 2000.

[5]"High Temperature Superconductivity, Bringing New Power to Electricity" news update from Bob Lawrence and Associates, Inc. had quoted 8,000 hours as of July 20, 2001, http://www.eren.doe.gov/superconductivity/pdfs/hts_update_july01.pdf (last accessed April 3, 2002).

[6]"High Temperature Superconductivity, Bringing New Power to Electricity," September 26, 2001.

[7]"High Temperature Superconductivity, Bringing New Power to Electricity," September 26, 2001.

project, IGC will install a one-quarter-mile second-generation HTS underground cable in Albany, N.Y.[8]

In addition, in May 2001, Copenhagen Energy began supplying electricity to its consumers through HTS electric cables from its Amager Substation. The new HTS cable is 30 m long, three-phase, and carries 2,000 A at 30 kV.[9]

Flywheel Energy Storage Systems

In this section, we describe the technical characteristics of flywheel energy storage systems, discuss the use of magnetic bearings, and provide data on the demonstration status of a flywheel energy storage system that uses HTS magnetic bearings.

Technical Characteristics

At its core, a flywheel energy storage system consists of a spinning flywheel and a motor/generator. The spinning flywheel is attached to the rotor of the motor/generator to enable the back-and-forth conversion of kinetic energy and mechanical energy. The motor/generator acts either as a motor when it stores electrical energy in the form of the kinetic energy of the rotor or as a generator when it converts the kinetic energy of the spinning rotor to electricity. When storing energy, the flywheel acts like a motor. A current flowing to the motor causes the rotor to spin faster so that electrical energy is changed into kinetic energy. The current must be AC and its frequency depends on the speed of rotation of the flywheel rotor. When providing energy, the flywheel acts like a generator. The spinning flywheel causes a current to flow and slows as some of the kinetic energy is changed into electrical energy. The generated current is AC and its frequency depends on the speed of rotation of the flywheel.

The frequency of the current supplied to the flywheel or generated by it varies with the rotor's speed of rotation. Generally, the frequency of the current supplied to or by the flywheel will not equal the frequency that a user is able to supply or needs in order to power equipment. As a result, a power electronics module is necessary to change the frequency of the current.

---

[8]See http://www.igc.com.
[9]"High Temperature Superconductivity, Bringing New Power to Electricity," September 5, 2001.

94

The written-pole motor[10] provides an alternative to using power electronics. The number of poles in this motor is not fixed. It behaves like an induction motor at low speed and like a synchronous motor at high speed. A motor/generator set can be built using the same principle and its output frequency is independent of its speed of rotation. A written-pole motor operating at 60 Hz has a maximum speed of 3,600 rpm. FESSs with rotors spinning at 3,600 RPM have been developed that rely on the written-pole motor for their operation. Most FESS developers, however, have designed their flywheels such that the rotor spins at speeds well in excess of 10,000 rpm. This arrangement requires power electronics for frequency conversion.

The kinetic energy stored in a flywheel is proportional to the square of the rotor's speed of rotation. According to the specifications of today's commercial and pre-commercial flywheel storage systems, these devices are used to provide power for relatively short periods of time, ranging from a few seconds to a few minutes. Some flywheels rely on mechanical rotor bearings that incur frictional losses. The substitution of magnetic bearings decreases these losses, but traditional magnetic bearings are mechanically unstable and require additional energy to control the rotor position. The advent of HTS bearings improved the performance of flywheels by obviating the need for these control mechanisms. The development of flywheels using HTS bearings allows flywheel energy-storage devices to operate with lower losses than those of existing systems.

The instability of magnetic bearings is a result of the forces that they experience. At equilibrium, the forces on an object add to zero. In addition, a stable equilibrium position is one in which a local minimum in potential energy exists. Because the potential energy everywhere around this position must be higher, displacing an object from the equilibrium position will cause it to experience a restoring force. This is illustrated by Figure A.1, which shows potential energy along the vertical direction as a function of position in the horizontal direction in a field in which stable equilibrium can occur. An object in this field will fall to the bottom of the well and remain there.

---

[10]Hoffman, Steve, "The Written-Pole Revolution," *EPRI Journal,* May/June 1997.

RAND*MR1531-A.1*

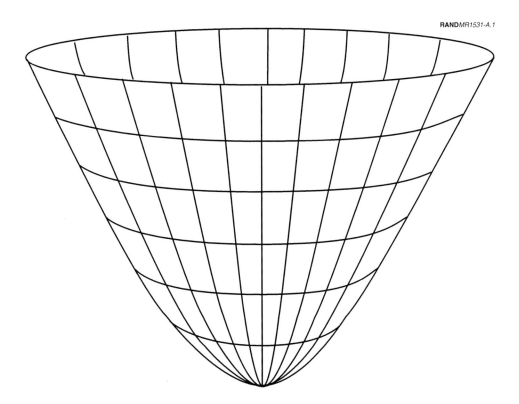

Figure A.1—Stable Potential Energy Distribution

According to Earnshaw's theorem,[11] a stable equilibrium cannot be achieved in a static magnetic field. This is a consequence of a fundamental property of magnetism—magnets always contain opposed poles (i.e., there are no magnetic monopoles). The potential energy for such a configuration is described by a "saddle point," which indicates that a local minimum in energy does not exist. This situation is illustrated by Figure A.2. An object will not be in a stable position anywhere in the potential energy field illustrated in the figure.

As a consequence of Earnshaw's theorem, any apparatus used to stabilize in space a configuration of magnets (excluding diamagnets, as discussed later) requires energy. Two types of magnetic bearings exist today: active and passive. Active magnetic bearings rely on sensors to determine the rotor position. This information is then used to feed a current to actuators, which move the rotor if necessary.[12] Passive magnetic bearings eliminate the need for sensors. Instead,

---

[11]Earnshaw, S., *Trans. Cambridge Phil. Soc.*, Vol. 7, 1842, p. 97, as quoted by E. H. Brandt in *Science*, Vol. 243, 1989, p. 349.

[12]Revolve Magnetic Bearings, Inc., *Magnetic Bearing Systems Basic Principles of Operation*, http://www.revolve.com/Basics.html, and *How Do Magnetic Bearings Work?*, http://www.revolve.com/Work.html (last accessed September, 25, 2000).

96

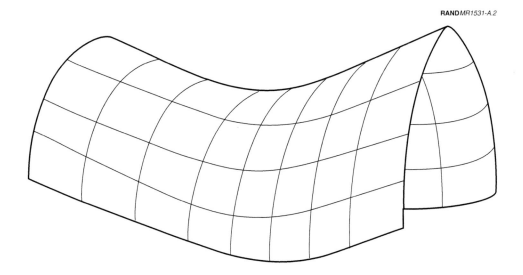

Figure A.2—Unstable Potential Energy Distribution

the spinning rotor induces currents in the inductive circuit elements surrounding it. These currents are used to adjust the position of the rotor by creating an external magnetic field.[13] Commercial flywheel storage systems rely on active magnetic bearings. Flywheels that rely on passive magnetic bearings are still under development.[14]

Earnshaw's theorem does not apply to diamagnets. In the presence of an external magnetic field, a diamagnet forms magnetic dipoles that oppose the field, and the diamagnet is consequently repelled by an external magnetic field. The strength of the diamagnet's repulsion depends on the strength of the applied magnetic field. This property has been utilized to make ordinary objects float but usually requires extremely large fields to be useful.

Superconductors are perfect diamagnets that exclude the magnetic field around them due to the Meissner effect.[15] A flywheel using HTS bearings thus does not require external stabilization of its rotor. Instead, the diamagnetism of the HTS

---

[13]Post, Richard F., *Dynamical Stable Magnetic Suspension/Bearing System,* United States Patent 5,9495,221, February 27, 1996, and *Passive Magnetic Bearing Element with Minimal Power Losses,* United States Patent 5,847,480, December 8, 1998.

[14]*Passive Magnetic Bearings 1999,* The Group of Passive Magnetic Bearings, the Institute of Robotics, the Swiss Federal Institute of Technology, Lausanne, Switzerland, http://www.epfl.ch/~jbermude/Gpmb/descrp.html (last accessed September 19, 2000); and Bowler, Michael E., "Flywheel Energy Systems: Current Status and Future Prospects," *Magnetic Material Producers Association Joint Users Conference,* September 22–23, 1997, pp. 1–9.

[15]Sheahen, Thomas P., *Introduction to High-Temperature Superconductivity,* New York: Plenum, 1994. The HTS materials are Type II superconductors, which allow some magnetic flux penetration as a function of temperature and magnetic field, so that these become design parameters for HTS magnetic bearings.

bearing maintains the position of the rotor relative to an external magnetic field. Because a feedback system is no longer needed to stabilize the rotor, the design of an HTS magnetic bearing is relatively straightforward compared with that of an actively controlled magnetic bearing.

Demonstration Status

Development of an energy storage system using HTS bearings, or an HTS FESS, continues. The team consists of Boeing Corporation, Praxair Specialty Ceramics, Ashman Technologies, Mesoscopic Devices (a Boulder Cryogenics subsidiary), Southern California Edison, and Argonne National Laboratory. Applications for this HTS FESS include load-leveling and power-quality management. The choice of application dictates the energy and power capabilities of the HTS FESS.

The energy storage capability that is envisioned for the HTS FESS is 10 kWh. The power output is either 3 kW for the load-leveling version or 100 kW for the power-quality management version. The first unit to be developed will have the 3 kW motor/generator; the second will have the 100 kW motor/generator. A charge and discharge cycle should return 88 percent of the energy supplied initially. Selected parameters of the flywheel are presented in Table A.3.

The HTS FESS has not yet been completed. The design of the 3 kW, 10 kWh system has been completed. The superconducting bearings, motor/generator, and control system have been constructed and are undergoing testing. Rotor construction is underway.[16] In addition, progress continues in the development and design steps for the 100 kW system. For FY 2002, the design of the motor/generator will be further refined.[17]

Table A.3

Selected Flywheel Parameters for the Boeing HTS FESS

Dimension	Size
Outside diameter	33.36 in.
Flywheel thickness	5 in.
Height of enclosure	8 in.
Weight	330 lb.

---

[16]*Superconductivity Partnership Initiative, Flywheel Electricity System, Project Fact Sheet*, July 31, 2001, U.S. Department of Energy, Washington, D.C.

[17]Hull, J. R., and A. Day, Development of Flywheel Energy System, *2001 Annual Peer Review Conference (SUP), DOE/EE-SESAPR/2001-CD*, available from the U.S. Department of Energy Office of Power Technologies, Washington, D.C.

HTS Transformers

In this section, we briefly describe the technical characteristics of HTS and conventional transformers and provide data on the demonstration status of HTS transformers.

Technical Characteristics

Each phase of a transformer consists of a primary coil and a secondary coil. The primary coil is the input (i.e., power flows into the transformer at this location). As current flows through the primary coil, it creates a magnetic field. If the flowing current is AC, the magnetic field will oscillate. The magnetic field passes through the secondary coil, where it induces an oscillating current according to Faraday's Law.[18]

An iron core may be present in the transformer to improve its operation by channeling the magnetic flux from the primary to the secondary coil. The induced current is the output. By varying the number of loops in the primary and secondary coil, the transformer can either increase or decrease the output voltage relative to the input voltage. Because the power (the product of the current and the voltage) flowing through the primary and secondary coils must be equal, the current will vary inversely with the voltage.

The electrical conversion of energy in transformers is not perfect. Energy losses result from a number of mechanisms, including Ohmic losses, Eddy current losses, induction losses, and hysteresis losses. Ohmic heating occurs because current flowing through the coils causes them to heat according to Ohm's Law. Eddy current losses take place because the changing magnetic field will induce currents everywhere in the transformer, including outside the coils. These currents dissipate by releasing heat. Hysteresis losses result in the metal core of the transformer because the core's magnetization does not exactly follow changes in its magnetic field.

During operation, a transformer heats up and this heat must be dissipated. In conventional transformers, this is achieved by immersing the coils in oil, which also serves to insulate them electrically. The hot oil conducts the heat away from the coils and prevents the transformer from overheating. In an HTS transformer, some heating also occurs and the cryogenic fluid conducts this heat away from the coils.

---

[18]Jackson, John David, *Classical Electrodynamics*, New York: John Wiley & Sons, Inc., 1962, p. 170.

In a conventional transformer, the core is immersed in the oil along with the coils. In an HTS transformer, however, the coils operate at cryogenic temperature whereas the core operates at a temperature somewhat above ambient. This approach avoids the penalty of removing heat generated by core losses at cryogenic temperature.

Demonstration Status

A number of prototype HTS transformers that are smaller and lighter than their traditional counterparts have been built and tested. Waukesha Electric Systems has built a 1 MVA transformer with a 13,800 V primary coil. The company also built a 5 MVA transformer with a 10 MVA overload capability with a 26.4 kV primary coil. In both cases, the transformers were built as prototypes to test the design for a 30 MVA transformer with an 11,000 V primary coil that Waukesha envisions as a commercially viable product. The full-size transformer is expected to use 200 pounds of HTS cable and weigh 24 tons compared with a weight of 48 tons for an equivalent copper-based transformer.[19] Waukesha, IGC, Oak Ridge National Laboratory, Rensselaer Polytechnic Institute, and Rochester Gas and Electric are all involved in this effort.

The 1 MVA transformer, tested in 1998, was used to test the general design approach for the HTS transformer. The 1 MVA single-phase transformer is a test bed with a core cross-section of a 30 MVA transformer with a 138 kV/13.8 kV rating.[20] By decreasing the temperature of the HTS windings, it was possible to operate the transformer at 1.65 MVA, compared with a design in which the transformer operates at 1 MVA.[21] Representative parameters of the 1 MVA transformer core and coil assembly design are listed in Table A.4.

Another effort—to develop a 10 MVA transformer—which was partially sponsored by the DOE was carried out by ABB, American Superconductor, Los Alamos National Laboratory, Southern California Edison, and American Electric Power. This effort was eventually discontinued.[22] Initially, ABB envisioned

---

[19]"World's Most Powerful HTS Transformer (HTS) Successfully Tested," http://www.waukeshaelectric.com/weshtspr.htm (last accessed August 31, 2000).

[20]Mehta, Sam P., Nicola Aversa, and Michael S. Walker, "Transforming Transformers," *IEEE Spectrum*, July 1997.

[21]"World's Most Powerful HTS Transformer (HTS) Successfully Tested," http://www.waukeshaelectric.com/weshtspr.htm (last accessed August 31, 2000).

[22]Baldwin, Thomas, "Center for Advanced Power Systems," CAPS 2001 Workshop, July 31, 2001.

100

Table A.4

Representative Parameters of the Waukesha 1 MVA Transformer Core and Coil Assembly Design

Parameter	Specification
RMS voltage and current	13.8/6.9 kV, 72.5/145 A
Cold mass weight	~ 590 kg (1,300 lbs.)
Vacuum tank volume	21,600 liters
Vacuum tank dimensions (h x w x d)	3.79 m x 2.78 m x 2.05 m
Vacuum tank weight	8,774 kg (19,300 lbs.)
Liquid nitrogen tank capacity	300 liters
Weight of liquid nitrogen tank and shield	364 kg (800 lbs.)
Cryocooler rating (Cryomech GB-37)	30 W at 25 K
Core weight	8,000 kg (17,600 lbs.)

SOURCE: "Selected High-Temperature Superconducting Electric Power Products," Superconductivity Program for Electric Systems, Office of Power Technologies, U.S. Department of Energy, January 2000.

developing a 100 MVA transformer operating between 225 kV and 20 kV. The 10 MVA transformer was to serve as the prototype to validate the design and approach.[23] This project was subsequently abandoned because the expected return on investment could not be met based on the present cost of HTS wire.

ABB initially built a transformer with a capacity of 630 kVA in an international effort with the Swiss federal government. Based on this experience, the company estimated a reduction in weight in the range of 20–35 percent and a reduction in volume in the range of 10–35 percent as compared with a conventional transformer.[24] Eventually, ABB plans to create a 100 MVA transformer. Key parameters describing this transformer are shown in Table A.5. Based on existing estimates, the transformer will be smaller, lighter, and more efficient than traditional transformers.

The performance of the 630 kVA prototype transformer was encouraging. Even so, ABB decided not to immediately pursue the market for HTS transformers based on a cost analysis. ABB concluded that the near-term market is still too small to justify the continued development of HTS transformers. ABB's cost-benefit analysis was based upon the assumption that investors in electricity transmission and distribution equipment (primarily utilities) expect the payback time to decrease from the 20 years originally proposed to 5 years. This makes the savings from HTS transformers less attractive than originally estimated. Meeting

---

[23]"Development and Demonstration of a 10 MVA High Temperature Superconducting Transformer," Annual Peer Review 2000, ABB, July 17–19, 2000; and *U.S. Department of Energy Superconductivity Program for Electric Systems*, proceedings on CD available from Energetics, Inc., Columbia, Md., order number DE00759424.

[24]Mehta, Sam P., Nicola Aversa, and Michael S. Walker, "Transforming Transformers," *IEEE Spectrum*, July 1997.

Table A.5

Summary of Key Parameters for the ABB HTS 100 MVA Transformer

Power	100 MVA
Voltage	225/20 kV
Cooling	Closed cycle
Volume savings	5%
Weight savings	20%
Load loss savings	80%
Up-front cost	150%
Life-cycle cost	90%

SOURCE: *Development and Demonstration of a 10 MVA High Temperature Superconducting Transformer; Phase I: HTST Benefits Evaluation,* ABB, 2000.

the new payback goals calls for prices of HTS wire to fall to $50 per kAm,[25] similar to the results of the analysis described in Chapter 5. In anticipation of the development of the next generation of HTS wire, the company chose to wait and re-enter the market at a more opportune time.

International efforts to develop and demonstrate HTS transformers have also been reported. The Swiss Federal Office of Energy Management sponsored work by ABB, Electricité de France, and Industriels de Genève to demonstrate a 630 kVA transformer in 1997.[26] In Japan, Fuji Electric, Sumitomo Electric, and Kyushu University built and tested a 500 kVA HTS transformer in 1995.[27]

The development of HTS transformers continues. A transformer for an HTS substation is under development. The team consists of IGC-SuperPower, Waukesha Electric Systems, and Southern California Edison. This project would demonstrate a prototype utility-sized HTS transformer to convert electricity from 66 kV to 12 kV.[28]

---

[25]"HTST Benefits Evaluation; Development and Demonstration of a 10 MVA High Temperature Superconducting Transformer," Annual Peer Review 2000, ABB, July 17–19, 2000; and U.S. Department of Energy Superconductivity Program for Electric Systems, proceedings on CD available from Energetics, Inc., Columbia, Md., order number DE00759424.

[26]Baldwin, Thomas, "Center for Advanced Power Systems," CAPS 2001 Workshop, July 31, 2001.

[27]*HTS Transformer Development in Japan*, September 1997, http://www.itri.loyola.edu/scpa/03_07.htm (last accessed March 29, 2002).

[28]"Department of Energy Announces Major Effort to Use High Temperature Superconductivity," *DOE News*, U.S. Department of Energy, September 24, 2001, Release No. R-01-161, http://www.energy.gov/HQPress/releases01/seppr/pr01161_v.htm (last accessed March 29, 2002.)

B. Parametric Relationships for Energy Loss and Life-Cycle Cost Comparisons of Superconducting and Conventional Cables and Conductors

The flow of an AC current through a conductor results in conduction losses. The mechanism responsible for this loss differs in the conventional and superconducting cases. We are not concerned here with the actual mechanisms, but only with the losses incurred by the different types of conductors. The conduction loss heats the conductor or cable and is proportional to its length, according to

$$P_c = \omega l,$$ (B.1)

where P_c is the power dissipated by conduction losses, ω is the conduction loss per unit length, and l is the length of the conductor or cable.

One critical difference between the superconducting and conventional cases is that the superconducting cable must be cooled to be operated. The heat generated by the ACSR conductor or XLPE cable is dissipated by the surrounding medium, whereas heat generated in the superconductor must be actively removed to maintain the superconductor's operating temperature.[1] In addition to conduction losses and the associated heating of the superconductor, the superconducting cables also incur thermal losses from heat leakage through the cable thermal insulation and the terminations at which the superconducting cable must connect to system elements at ambient temperature.

Based on the power-loss mechanisms we just described, thermal losses for a superconducting cable are given by

$$P_t = (\theta + \omega +)lt \quad ,$$ (B.2)

---

[1]Some underground cables are actively cooled by flowing oil that must be pumped through the system. The energy requirements and cost of such systems are not included in our analysis. Overhead transmission line maximum current ratings are limited by the ambient cooling determined by temperature and wind speed.

104

where P_t is the power dissipated by thermal losses, θ is the thermal loss per unit length of the cable, ω is the electrical loss per unit length of the cable, l is the length of the cable, and τ is the thermal loss at the cable terminations.

The superconducting cables must be cooled to maintain their operating temperatures of 65 K for the HTS cable and 8 K for the LTS cable. We next calculate the power needed to keep the superconducting cables at their respective operating temperatures. Equation (B.2) determines the heat load that must be removed by the cooling system. According to the second law of thermodynamics, the best performance that can be achieved by any refrigerator operating between temperatures T_h and T_c, expressed in degrees Kelvin, is the Carnot efficiency,[2] η_c, which is given by

$$\eta_C = T_c \: / \: (T_h - T_c).\tag{B.3}$$

where T_c is 65 K for the HTS cable, T_c is 8 K for the LTS cable, and T_h is 298 K (25°C) for both HTS and LTS cables. From Equation (B.3), we see that for the HTS cable the best possible efficiency is slightly less than 28 percent, and for the LTS cable the best possible efficiency is slightly less than 3 percent. In other words, even with perfect refrigeration technology, 3.6 watts of power must be supplied for each watt of heat load to maintain the HTS cable at its 65 K operating temperature, and 36.25 watts of power must be supplied for each watt of heat load to maintain the LTS cable at its operating temperature of 8 K. In fact, current cryocooler technology ranges from a few percent of Carnot efficiency to 20 percent of Carnot, with higher efficiencies for larger machines.[3] So in actual practice, the cooling power demands are several times greater. We define the coefficient of performance, ρ, as the amount of electric power in watts that must be supplied to the cooler to remove one watt of heat load at the superconducting cable operating temperature in terms of η, the efficiency of the cooler relative to its Carnot efficiency, by the following equation:

$$\rho = 1 / \eta \: \eta_c.\tag{B.4}$$

For the HTS cable, the commercial Gifford-McMahon-type AL300 cryocooler manufactured by Cryomech, Inc.,[4] requires 7,400 W of input power to remove 285 W of heat load at 65 K, which gives $\eta = 0.14$, and a number of Stirling-type and pulse-tube–type cryocoolers with 100–1,000 W heat removal capability at

___

----

[2]See, for example, Fermi, Enrico, *Thermodynamics*, New York: Dover Publications, Inc., 1936, Chapter III.

[3]Radebaugh, Ray, "Development of the Pulse Tube Refrigerator as an Efficient and Reliable Cryocooler," submitted to the Proceedings Institute of Refrigeration (London, England) 1999–2000, National Institute of Standards and Technology, Gaithersburg, Md., 2000.

[4]Specifications are available at http://www.cryomech.com.

80 K have achieved η between .15 and .20.[5] Thus, for the HTS cable, a realistic range for ρ based upon current technological capability is 18 to 26, so up to 26 W of power must be supplied to the refrigerator for each watt of heat load removed at 65 K. By comparison, for the LTS cable, an experimental system developed at Brookhaven National Laboratory required 250 W of refrigeration power per W of heat load operating at 6.7 to 8 K.[6] Thus, the lower operating temperature of the LTS cable brings with it an order of magnitude penalty in required refrigeration power.

The total losses in the cable are given by

$$P = P_r + P_c = \rho P_t + P_c, \tag{B.5}$$

where P_r is the refrigeration power and P is the total power lost in operating the superconducting cable. Conduction is counted twice: The first time it is counted is part of the thermal power dissipation of the cable, P_t, because conduction heats the cable, providing part of the cooling load; the second time it is counted because some of the power flowing through the cable, P_c, is dissipated in the form of conduction losses and this power cannot be transmitted by the cable. By contrast, the thermal and termination losses are only counted once, as part of P_t, because they arise from heat leaking into the cable. Thus, the overwhelming majority of the losses in the superconducting cable can be traced to the need for refrigeration.

We now derive an expression for the power dissipated in cables or conductors of a given length, which will allow the determination of cable or conductor length such that the superconducting cables dissipate less power during operation than the conventional technologies. From Equations (B.1) through (B.5), we find that for the superconducting cables

$$P = P_r + P_c = \rho[(\theta \omega + \psi l \omega] + \iota l, \tag{B.6}$$

and for the ACSR conductor and XLPE cable

$$P = P_c = \omega l. \tag{B.7}$$

We compare the superconducting and conventional technologies per unit of delivered power. Because the ACSR conductor and the XLPE cable carry less power than the superconducting cable, the comparison per unit of delivered

---

[5]Radebaugh, Ray, "Development of the Pulse Tube Refrigerator as an Efficient and Reliable Cryocooler," submitted to the Proceedings Institute of Refrigeration (London, England) 1999–2000, National Institute of Standards and Technology, Gaithersburg, Md., 2000.

[6] Gerhold, J., "Power Transmission" in B. Seeber, ed., *Handbook of Applied Superconductivity*, Volume 2, Bristol, UK: Institute of Physics Publishing, 1998.

power is accomplished by multiplying the power losses of the cables by the ratio of power carried, R.

We see from making Equation (B.6) equal to R times Equation (B.7) that the minimum length required for the superconducting cable to compete with the conventional technologies per unit of power delivered depends upon the magnitude of the thermal and termination losses, which contribute to the heat load, the efficiency of the cryocooler (which is incorporated in the coefficient of performance, p, and the power ratio, R, according to

$$l_{MIN} = \frac{p}{R\omega_c\, p\theta\, (\omega + \omega_s)_s},$$

(B.8)

where the subscript s signifies superconductor and the subscript c signifies conventional (ACSR) conductor or (XLPE) cable.

We note that when the denominator of Equation (B.8) is greater than zero, the length is positive and finite. This corresponds to the case in which the losses in the superconducting cable are lower than those in the conventional cable or conductor for any $l > l_{MIN}$. When the denominator is less than zero, the length is negative and finite. This is the case in which the losses in the conventional cable or conductor are lower than those in the superconducting cable for all lengths.

Setting the denominator of Equation (B.8) equal to zero provides a parametric relationship for the boundary between these two regions:

$$p\theta\, \omega_{s} + \omega\, R\omega_{c\,s}.$$

(B.9)

Equation (B.9) sets a lower limit on the performance of the cryocooler as a function of the specified parameters of the conventional conductor or cable and the conduction and thermal loss in the superconducting cable.

In the comparison to this point, we assumed that the cable or conductor was used continuously at its maximum rated power. In reality, a cable or conductor may carry a lower current than its full capacity, depending upon the operating conditions.

We define utilization as the fraction of the maximum rated power carried by the conductor or cable. Because the conduction loss is proportional to the current squared for the conventional conductor or cable and to the current cubed for the superconducting cable,[7] different time variations of utilization can lead to

---

[7] Gerhold, J., "Power Transmission" in B. Seeber, ed., *Handbook of Applied Superconductivity*, Volume 2, Bristol, UK: Institute of Physics Publishing, 1998.

different values of loss. It is customary to define a load curve that represents the variation in utilization over a given time period, such as a day, and then to integrate the appropriate function of the current using this curve to determine the loss. For the purpose of this analysis, we adopt the simplified procedure of assuming that the current flows at its maximum level for a fraction of the time and does not flow at all the rest of the time.[8]

We calculate the minimum length of superconducting cable required for it to compete with an ACSR conductor or XLPE cable per unit of power delivered based on different levels of utilization. As utilization decreases, the length of the superconducting cable must increase for the cable to remain competitive. This is because even when it does not carry current it must be cooled because the thermal and termination losses occur regardless of the current flow. For utilization less than 1, Equation (B.8) becomes[9]

$$l_{MIN} = \frac{p}{uR\omega_c \, p\theta \, (\text{+}\omega \, u \, \text{+}\theta) \, u \, _s}.$$ (B.10)

The variable u, which indicates the utilization, has a value between 0 and 1 and multiplies only the conduction losses. Thermal and termination losses do not depend on utilization.

If we set the denominator of Equation (B.10) to be equal to zero, we see that the utilization factor, u, enters the parametric relationship of Equation (B.9), according to

$$p\theta \, \text{+}\omega u \, _s \text{+} \omega u R \, \omega \, u \, _{s},$$ (B.11)

which is also Equation (4.1) in Chapter 4.

The acquisition cost of the cable or conductor, A_c, can be defined according to

$$A_c = l \, I \, \chi,$$ (B.12)

where l is the length in meters, I is the current in thousands of amperes, and χ is the cost in dollars per kA·m.

In the case of the superconducting cable, the cost of acquiring a cryocooler must be added to the cost of acquiring the cable. The acquisition cost of the cryocooler, or of refrigeration, A_r, equals

---

[8]Alternatively, one could use the instantaneous value of the current. The difference between the two is negligible at high utilization.

[9]If instantaneous value of the current were used, u^2 would appear in the ω_c term and u^3 would appear in the ω_s terms in Equation (B.10). The difference is negligible for high utilization (e.g., $u > 0.6$) and increases in importance as the utilization decreases.

108

$$A_r = r[(\theta + \omega + l)\tau \quad],$$ (B.13)

where r is the cost per watt of the cryocooler and the expression in brackets is the heat load, with θ, ω, l, and τ as defined earlier.

Next, we consider the cost of operating the cable or conductor. As discussed in the previous section, the action of passing a current through the cable or conductor results in a loss of power. This loss is proportional to length as shown in Equation (B.1). The cost resulting from this loss of power may be computed according to

$$O_c = 0.0876 \, \varepsilon \, d \, l \omega \quad ,$$ (B.14)

where O_c is the contribution of conduction losses to the operating cost of the cable or conductor, ε is the cost of electricity in cents per kilowatt hour, and d is the discount factor to change the cost of operating over a single year to the cost over the lifetime of the cable or conductor at the appropriate discount rate. We use a discount rate of 7 percent, for which $d = 14.3$ for the assumed 40-year operating lifetime.

In similar fashion, the cost of refrigeration for the superconducting cable may be computed using Equation (B.2) according to

$$O_r = 0.0876 \, \varepsilon \, d\rho [(\omega \quad \tau)l \quad],$$ (B.15)

where O_r is the contribution of refrigeration to the operating cost of the cable, ε and d are as defined earlier, ρ is the coefficient of performance, and θ, ω, and τ are the thermal, conduction, and termination losses, respectively, all of which were previously defined.

We based the minimum length for the superconducting cable to compete with the ACSR conductor or XLPE cable on energy considerations only. Now, we make the same comparison, but base it on life-cycle cost. By analogy with Equation (B.8), the minimum length at which the superconducting cable will have a lower life-cycle cost than the ACSR conductor or XLPE cable per unit of delivered power can be derived from Equation (B.12) through (B.15), according to

$$l_{MIN} = \frac{r\tau + 0.0876\varepsilon \, \rho d}{R I \chi_c - I \chi_s \, \theta(\omega +_s) \quad 0.0876 \, \omega d [\rho R \, \rho\theta - \omega_s] \quad (_s)]},$$ (B.16)

where, as in Equation (B.8), the subscript c denotes conventional and the subscript s denotes superconductor, and R is the power ratio defined previously. When the denominator of Equation (B.16) is positive, the superconducting cable will have a lower life-cycle cost than the ACSR conductor or the XLPE cable for all $l > l_{MIN}$, whereas if the denominator is negative, the superconducting cable will have a higher life-cycle cost than the ACSR conductor or XLPE cable for any length. The denominator of Equation (B.16) shows that the cryocooler efficiency requirement for a lower superconducting cable life-cycle cost depends on the relative cost of the superconducting cable and cryocooler and the ACSR conductor or XLPE cable. For example, if the cost of the cryocooler (excluding the cooling load associated with the cable terminations) plus the cost per unit of delivered power of the superconducting cable is less than the cost per unit of delivered power of the ACSR conductor or XLPE cable, so that

$$RI\chi_c - \chi_s = \theta_r(\omega_c > \omega_s) \quad 0, \tag{B.17}$$

then the efficiency of the cryocooler for the superconducting cable life-cycle cost benefit does not have to be as high as that for the superconducting cable energy benefit. However, if the inverse cost relationship holds, and the inequality of Equation (B.17) is reversed, then the cryocooler must be even more efficient for the superconducting cable life-cycle cost benefit than for the superconducting cable energy benefit.

Setting the denominator of Equation (B.16) to be equal to zero provides a parametric relationship analogous to Equation (B.9):

$$RI\chi_c - \chi_s = \theta_r(\omega_c + \omega_s) \quad 0.0876\omega_c d[\omega R_c \theta_r(\omega_c - \omega_s)] \quad (\omega_c \quad \omega_s)] \quad 0. \tag{B.18}$$

The minimum length calculated from Equation (B.16) is the length for which the life-cycle cost of acquiring and operating the superconducting cable and the conventional cable or conductor to which the superconducting cable is being compared are equal after 40 years. This corresponds to a 40-year payback time for the operational cost savings of the superconducting cable, as compared with the difference in acquisition cost between the superconducting cable plus the cryocooler and the conventional cable or conductor. Longer cables will provide payback in a shorter period of time. If we relax the 40-year payback time assumption, Equation (B.16) provides a relationship between length and the discount factor, d, for a life-cycle cost break-even between the HTS cable and the conventional cable or conductor. The discounted payback time can then be calculated from d using well-known formulas.[10] For cables with length much longer than the minimum length, the discounted payback approaches an

---

[10]See, for example, Weston, J. Fred, and Eugene F. Brigham, *Managerial Finance*, 3rd ed., Holt, Rinehart and Winston, 1969, pp. 179–182. The payback period often neglects the time value of money. As a result, many analyses rely on net present value calculations instead. In our analysis, the cash flows are constant from year to year and the net present value relation can be inverted to yield a discounted payback period calculation that takes into account the time value of money.

110

asymptotic value, which can be computed from the asymptotic value of d, given by

$$d_{ASYM} = \left(\frac{1}{0.0876\varepsilon}\right)\frac{I\chi_s + r(\theta + \omega_s)\chi RI_c}{[R\omega_c\omega\rho\theta\omega(+_s)]}, \tag{B.19}$$

where, as earlier, the subscript s denotes superconductor and the subscript c denotes conventional, and all parameters have been previously defined. One can solve Equation (B.16) for d and expand the solution in terms of $1/l$ to obtain an approximate solution for d that will be valid for cable lengths such that

$$l > \frac{p}{R\omega_c\omega\rho\theta\omega(+_s)} = l_{MIN(0)}, \tag{B.20}$$

where $l_{MIN(0)}$ is the minimum length for zero acquisition cost derived from energy considerations in Equation (B.8). The expression for d to lowest order in $1/l$ is given by

$$d = d_{ASYM} + \frac{r\tau[(R\omega_c\omega\rho\theta\omega(+p+\chi)]\chi[R\theta\omega I+_s r(_s)]}{0.0876\varepsilon[(R\omega_c\omega\rho\theta\omega(+_s)]^2}\left(\frac{1}{l}\right) + O\left(\frac{1}{l}\right)^2$$

$$\tag{B.21}$$

C. Parametric Relationships for Comparison of the Energy Use of Flywheel Energy Storage Systems and Battery Energy Storage Systems

The power flows in an energy storage device, which are illustrated in Figure 5.1 of Chapter 5, can be represented mathematically as

$$P_L = (1 - u)I_L + uP\left(\frac{1}{C_{r/i}C_s}\right),$$

(C.1)

where P_L is the total power dissipated by the system, I_L is the idling loss, u is the utilization defined as the fraction of time that energy is being drawn from the storage system, P is the power output, $C_{r/i}$ is the rectifier/inverter conversion efficiency, and C_s is the conversion efficiency of the energy storage module.

When comparing two energy storage devices, the simplest situation is one in which both devices have the same energy storage and power output. This is because some losses scale with the power output of the storage system whereas others scale with the amount of energy stored. For two systems with the same energy storage and power output, the losses can be compared directly and a parametric relationship can be defined among the different types of losses and system utilization. Analysis of this relationship then enables determination of the relative desirability of these particular FESSs and BESSs from the standpoint of energy use.

If the power output and energy storage of the two systems being compared are not the same, the devices should be combined so that their properties match. For example, if System A has three times the power output of and the same energy storage capability as System B, then three devices from System B should be combined in parallel to produce the same output as System A. Similarly, if System A has three times the energy storage capability of and the same power output as System B, then the three devices from System B should be combined in series to give the same output as System A. Losses in System B should be scaled according to how many devices are used.

After establishing that the two systems of interest have equivalent power output and energy storage characteristics, Equation (C.1) can be written for each system,

112

the two equations can be set equal to each other, and the resulting equation solved for utilization. The result is Equation (C.2), which provides a parametric relationship between the utilization and the system losses for the energy use of the FESS to equal that of the BESS:

$$u = \frac{\bar{I}_F}{\bar{I}_F - \left\{ \left(\dfrac{1}{C_{r/i}C_s} \right)_F - \left(\dfrac{1}{C_{r/i}C_s} \right)_B \right\}},$$ (C.2)

where $\bar{I}_F$ is the idling loss of the FESS per unit output power, the idling loss of the BESS is neglected, the subscript B refers to the BESS, the subscript F refers to the FESS, and all other definitions are as in Equation (C.1). For Equation (C.2) to be meaningful, the energy conversion efficiency of the FESS energy storage module $(C_s)_F$ must be higher than that of the BESS energy storage module $(C_s)_B$.

D. Parametric Relationships for Comparison of Energy Use and Life-Cycle Cost of HTS and Conventional Transformers

The losses in a conventional transformer can be expressed as

$$P = c + \omega, \tag{D.1}$$

where P is the power dissipated, c is the core loss, and ω is the conduction loss.

The HTS transformer core is not typically operated at cryogenic temperature while the coils are maintained at cryogenic temperature. Thus, the heat load to be removed by the cryocooler includes the coil conduction losses and the thermal heat leakage through the insulation and the terminations but not the core loss. Therefore, the power dissipated by the HTS transformer can be expressed as

$$P = c + \rho(\theta + \omega), \tag{D.2}$$

where the variables are as in Equation (D.1), with the addition that ρ is the cryocooler coefficient of performance and θ is the thermal loss, both previously defined in Appendix B. The factor ρ depends on the operating temperature and the efficiency of the cryocooler according to Equation (B.4).

Equations (D.1) and (D.2) assume that the transformer operates at its maximum rating at all times. However, electric demand varies over the course of a day so that it is necessary to modify the loss equations to take utilization into account. Using the same approximation for utilization as was described in Appendix B, the total power loss in the HTS transformer can be written as

$$P = c + (\theta + u\omega)\rho, \tag{D.3}$$

where all the terms are the same as in Equation (D.2) with the addition that u is the utilization. Similarly, Equation (D.1) for the conventional transformer becomes

$$P = c + u\omega. \tag{D.4}$$

If we set Equation (D.4) equal to Equation (D.3), we obtain the following parametric relationship between utilization and cryocooler efficiency that

114

determines when the HTS transformer will dissipate the same amount of power as the conventional transformer:

$$c_c + u\omega_c = t_s (t_s \omega p_s) ,$$ (D.5)

where all variables are the same as in Equations (D.3) and (D.4) and the subscript c denotes conventional, and the subscript s represents superconductor.

The life-cycle cost of a transformer is the sum of its acquisition cost and of the yearly operating cost multiplied by the discount factor that takes into account the discount rate and time period of interest. By analogy with Equations (B.12) through (B.15) in Appendix B, the life-cycle cost of a transformer can be represented as

$$LCC = r + A_t \quad 0.0876\varepsilon \, dP ,$$ (D.6)

where LCC is life-cycle cost, r is the acquisition cost of the cryocooler, A_t is the acquisition cost of the transformer, 0.0876 converts the cost of electricity in cents per kilowatt hour to cost of electricity per year in dollars, ε is the cost of electricity in cents per kWh, d is the discount factor, and P is the amount of power dissipated in kW, which is given by Equation (D.3) or (D.4).

Equation (D.6) can be used to derive a parametric relationship between the HTS acquisition cost, the cryocooler cost and efficiency, and the cost of electricity in order for the HTS transformer life-cycle cost to equal that of the conventional transformer. When evaluating the life-cycle cost of the conventional transformer, $r = 0$ in Equation (D.6). Alternatively, one can treat d as a variable and compute payback times for given values of the parameters, following the same procedure that led to Equation (B.20) in Appendix B.